WK 547

Giant Molecules

Giant
Molecules

From Nylon to Nanotubes

WALTER GRATZER

OXFORD
UNIVERSITY PRESS

OXFORD
UNIVERSITY PRESS

Great Clarendon Street, Oxford OX2 6DP

Oxford University Press is a department of the University of Oxford.
It furthers the University's objective of excellence in research, scholarship,
and education by publishing worldwide in

Oxford New York

Auckland Cape Town Dar es Salaam Hong Kong Karachi
Kuala Lumpur Madrid Melbourne Mexico City Nairobi
New Delhi Shanghai Taipei Toronto

With offices in

Argentina Austria Brazil Chile Czech Republic France Greece
Guatemala Hungary Italy Japan Poland Portugal Singapore
South Korea Switzerland Thailand Turkey Ukraine Vietnam

Oxford is a registered trademark of Oxford University Press
in the UK and in certain other countries

Published in the United States
by Oxford University Press Inc., New York

British Library Cataloguing in Publication Data

Data available

Library of Congress Cataloging in Publication Data

Data available

Typeset by SPI Publisher Services, Pondicherry, India
Printed in Great Britain
on acid-free paper by
Clays Ltd, St Ives plc

ISBN 978–0–19–955002–9

1 3 5 7 9 10 8 6 4 2

Contents

List of Illustrations

Preface

Giant molecules, or macromolecules, or polymers, do many indispensable things for us: they allow us to see, hear, breathe, move, and fend off viruses and bacteria; they make up the solid parts of our bodies, and one of the most famous of them harbours the genes that we transmit from generation to generation, They feed us, and provide us with the means to clothe and shelter ourselves. Paper is made of polymers, and so are paints, adhesives, and packaging. Without polymers our cars would not run, nor aeroplanes fly. And yet their existence was hidden from science until little more than a century ago, and remained for some time after that a matter of hot, often rancorous, controversy. By the early years of the twentieth century, though, the evidence ought to have seemed indisputable. Yet even three decades on there were still a few renegade scientists in reputable centres of learning stridently insisting that molecules with more than a few hundred atoms did not, and could not, exist. Such denial required an obdurate refusal to confront facts, but as the economist J. K. Galbraith noted, faced with the options of changing their minds or proving that there was no need to do so, most people get busy with the proof.

Man-made polymers emerged from a halting start at the dawn of the past century as products of a huge worldwide industry, where by mid-century they had become inseparable from almost every sphere of human activity. Today there is a great ebullition of new materials, which are being integrated into most of the appurtenances of modern life, from bullet-proof jackets, surrogate skin, adhesives that work under water, dirt-repellent coatings, and surgical implants to flashing diodes, liquid crystal displays, and electronic circuits so small as to be visible only under the microscope. Many of the most remarkable contrivances are modelled on materials found in nature, for evolution has been generating innovations for some 3 billion years, and has taught scientists many lessons. The properties of DNA, for example, have no parallel in any materials conceived by man, not only in its capacity to store information, in which it far exceeds that of the most advanced computer, but also in the versatility of the submicroscopic structures to which it can give rise—nano-scale robots, perhaps, of the future. Lessons learned from the chemistry and physics of natural giant molecules have inspired the design of 'smart' materials—coatings and clothes, amongst many others, which can respond to changes in temperature, humidity, or light levels.

And then there are the processes of life itself, over which we now have a measure of control, for we can manipulate the genes of plants and animals, and produce in bacteria and plants an infinite range of proteins for medical and other uses. Insulin and other hormones are now produced in bacteria, thereby eliminating the hazards of infection by pathogens from animal sources. We search through DNA to discern hereditary abnormalities and thus specify treatments for the diseases they engender, or to solve crimes and resolve paternity disputes. Soon we may all carry in a locket the encrypted readout of our genetic make-up to ensure the best treatment for malfunctions and mishaps.

The science of polymers, natural and artificial, is then a subject of enormous scope, and a book the size of this can do it scant justice. I hope all the same that this brief survey of the field and its often colourful history will convey its allure, and perhaps even stimulate further explorations. The narrative demands no advanced or specialized knowledge, and is meant to be accessible to the layman. Chemistry at a basic level would of course be helpful, and I have laced the text with formulae, mainly of simple compounds, to prod the memories of readers who have forgotten what they may have learned at school. I have provided a list of books and articles for anyone wishing to pursue the subject in greater depth.

I am grateful for the encouragement of my editor and friend, Latha Menon, and to Emma Marchant and Fiona Vlemmiks at Oxford University Press, to Fiona Orbell, who searched for illustrations, Charles Lander for expert copy-editing, and Kate Kirwen, whose skills in creating most of the diagrams were indispensable.

1

BEFORE THE DAWN

Molecules perceived

By the middle of the nineteenth century the lineaments of chemistry had been well enough defined. Many of the chemical elements of which all matter is made, and all of the most abundant ones, had been identified, and their *atomic weights* had been determined. These atomic weights were expressed relative to that of the lightest element, hydrogen, which was allocated a weight of 1. On this basis the weight of carbon was 12, nitrogen was 14, oxygen was 16, sulphur was 32, and so on. It was recognized that atoms of these elements combined with each other in defined proportions to make molecules, and that each element had a characteristic *valence*; its atoms, in other words, could unite with a fixed number of other atoms of the same or another kind. So hydrogen had a valence of 1—could bind to only one other atom—while oxygen had a valence of 2, carbon (the key element in macromolecules) of

4, nitrogen of 3, and so on. As time went on and analytical methods evolved, the formulae of familiar compounds were uncovered. Water molecules were found to consist of two hydrogen atoms and one oxygen, or in modern parlance, its formula is H_2O, and its *molecular weight** therefore $16 + 2 = 18$; sulphuric acid, or oil of vitriol as it used to be called, contains a sulphur atom, and was H_2SO_4 (molecular weight $2 \times 1 + 32 + 16 \times 4 = 98$), while acetic acid, the acid of vinegar (*acetum* in Latin), is $C_2H_4O_2$. Acetic acid, a product of bacterial fermentation, is a compound of carbon, as indeed are practically all substances produced by living organisms.

In 1700, when acetic acid was first purified, and for the next century, science was pervaded by vitalism, the doctrine that set apart the products of living organisms from all other matter. These products were permeated, the vitalists believed, by a 'vital force', and therefore such compounds could be generated *only* by living organisms, and could never be synthesized in the laboratory. This article of faith was proved false in 1828, when a German chemist, Friedrich Wöhler, effected a laboratory synthesis of urea, the metabolic waste product expelled in our urine. But it was

---

* To satisfy the pedants it must be said that the term, which served chemists perfectly well for two centuries, has been replaced by *molecular mass* with units of daltons (abbreviated Da), in honour of the reticent Manchester Quaker John Dalton, who, in *A New System of Chemical Philosophy* (1808), presented evidence that the common elements combined with one another in fixed weight ratios. This implied that the elements consisted of defined packets; in short: atoms. One dalton is the mass of a hydrogen atom, or strictly speaking one-sixteenth the mass of an oxygen atom (which is infinitesimally different). In common with almost universal practice amongst chemists we will cling here to the forbidden usage. The reason for switching to the oxygen standard is that the elements are mixtures of *isotopes*. Hydrogen contains a tiny proportion of atoms of atomic weight 2; this is 'heavy hydrogen', or deuterium, written H^2. Carbon contains a similarly minute proportion of C^{13}. So the standard is one-sixteenth of the weight of the vastly preponderant oxygen isotope, O^{16}.

also by then becoming clear that carbon had a unique position among the elements, for unlike all the others it could give rise to a limitless variety of compounds, large and small. This, as will be seen, is due to its propensity to combine with itself and form chains of connected atoms, to which other elements can attach themselves. That special attribute of carbon engendered a self-contained branch of chemistry, designated by the misnomer *organic chemistry*, thereby distinguishing itself from the chemistry of all the other ninety-one natural elements (plus a few more, created in high-energy machines by physicists), known as *inorganic* chemistry. There is one exception to the general rule: silicon, Si, an element belonging to the same chemical group as carbon, possesses in part the property of forming long chains, though only in alternation with oxygen. This is the essence of ordinary glass. Compared to the chemistry of carbon, though, that of silicon is therefore an impoverished thing, with the number of compounds known to organic chemistry now standing at a stupefying 16 million or so, of which about 2 million have been synthesized in the laboratory.

The colloid debates

But there is also *physical* chemistry, the branch of chemistry that treats of the physical nature and behaviour of molecules, and it was from this direction that the idea of giant molecules emerged, at least so far as chemists were concerned, and not without decades of acrimonious debate. In 1861 a noted English chemist, Thomas Graham, was studying the rate at which the molecules of various compounds diffused through the pores of membranes, made of such materials as parchment. Most substances passed through easily, but there were a few that escaped very slowly or not at all. Among them were starch, gelatin, and the substance recovered from blood and other animal fluids: albumen. Their passage was, Graham concluded, impeded due to their large size. He supposed that they consisted of clusters, or aggregates, of the kinds of molecules, all of them small, then known to chemists. Because solutions of starch and gelatin and like substances are viscous and sticky, Graham coined the term *colloid*, from the Greek for glue. Another term for such aggregates, which came into vogue around this time, was *micelle*. Today the expression is almost exclusively applied to detergent molecules, which have the property of clustering together when dissolved in water (allowing them to take up greasy materials in the fatty centre of the cluster), but then it was synonymous with colloid.

Figure 1 Gerhardus Johannes Mulder (1802–80), the Dutch chemist who determined the identity and proportions of the elements present in proteins.

Not all colloids were organic substances; dispersions of silica gel (SiO_2), for instance, were also held back by Graham's membranes. (Indeed, a century later it was discovered that metals such as gold could likewise be converted into a seemingly soluble, colloidal state.) Some researchers saw no good reason to doubt that starch and proteins, such as gelatin and albumin, were simply very large molecules, but the majority seem to have sided with Graham: there was no such thing, they held, as a giant molecule, only large sticky aggregates.

And yet the evidence for giant molecules, or let us call them macromolecules, had been in existence for twenty-five years: in 1838 there had appeared a remarkable paper by a doctor, chemist and physiologist, Professor at the University of Utrecht in the Netherlands, Gerhardus Mulder. Mulder had applied the methods of elemental analysis—determining the proportions of atoms of different elements in a compound—to albumin from blood serum and from egg-white, and to fibrin from blood. He found that all three of these substances—for which his patron, the great Swedish chemist Jöns Jacob Berzelius, coined the collective noun 'protein' (taken rather opaquely from the Greek for 'primary')— had closely similar compositions. They were made up mainly of carbon, nitrogen, oxygen, and hydrogen, but there were also small amounts of sulphur and of phosphorus.

All this led Mulder to the formula $C_{400}H_{620}N_{100}O_{120}P_1S_1$ for two of his proteins, and the same for the third, blood serum albumin, but with one more sulphur atom. Now, since no compound can contain less than one atom of any element (unless it is not there at all), this sets a lower limit for the number of atoms in a molecule, and thus for its minimum molecular weight, which for Mulder's three proteins comes out at about 14,400. It is the minimum molecular weight, because the analysis delivers only the *proportions* of the constituent elements, the true value possibly being 28,800 or 43,200 and so on—assuredly, in any event, a macromolecule. Mulder later modified his view that sulphur and phosphorus were integral parts of the protein molecule: they were, he decided, impurities, and in this he was correct about phosphorus, but wrong about sulphur. A ferocious dispute now erupted with the influential and overbearing German chemist Justus von Liebig, who sought to persuade the world that Mulder's analyses were worthless. Indeed, if there was no single sulphur atom the logic of the reasoning leading to the high molecular weight would collapse, leaving the simple formula (preserving the ratios of the four elements) $C_{20}H_{31}N_5O_6$. This would still amount to a pretty big molecule, but the precision of the analyses was not such that the H_{31} could with any confidence be discriminated from H_{30}, and in that case how could the simple composition C_4H_6NO be ruled out?

Yet Mulder clung to his conclusion—14,400 was the molecular weight of the *Grundstoff*, the core substance, of proteins—and traded insults with Liebig, in a tone unthinkable in the academic discourse of today. (Liebig had the upper hand, for not only was he a master of sarcastic obloquy, but he also had his own resource for its dissemination: he was both founder and editor of an important journal, *Liebig's Annalen der Chemie*, and made full use of its columns in hounding his rivals.)

The issue remained in doubt until the red oxygen-carrying protein of the blood, haemoglobin, was crystallized in 1840. This ran counter to the German organic chemists' view of the *Schmieren*—the despised smears or pastes yielded up by biological materials—for it was held, with some reason, that crystals were made up solely of identical molecules in the pure state. (In fact Graham coined the now-antiquated term *crystalloid* for any compound that was not a colloid.) Haemoglobin, moreover, had been shown to contain iron (hence its red colour), and however rigorous the purification process, the content of iron never deviated from 0.4%, leading to a minimum molecular weight for the protein of 16,000, an uncannily good estimate, for the right answer is 16,700.

This, though, was still not enough to convince the implacable aficionados of the colloid doctrine, or as their opponents often called them, the micellists. Around the turn of the twentieth century two German organic chemists, Emil Fischer and Franz Hofmeister, established, simultaneously and independently, the general chemical nature of proteins (which we shall come to presently). Fischer also developed methods of chemical synthesis of protein chains from their building blocks, amino acids. He managed to assemble a chain of eighteen of these units—a tour de force at the time. This 'proto-protein' (which we would now call an oligopeptide, from the Greek *oligos*, a few, as in oligarchy) would have had a molecular weight of about 2,000, which Fischer, initially at least, believed must represent the size of natural proteins. This is of course a gross underestimate, for in reality there are proteins a thousand times larger. There was still, then, no bridge over the chasm separating the organic from the physiological chemists, the *Schmierchemiker*.

There was according to a common belief, an upper limit of size, above which a molecule would simply fall apart. The organic chemists of the day, accustomed to dealing with molecules made up of tens of atoms, could not accommodate their minds to an actual chemical compound as vast as the biochemists were proposing. Yet even in Mulder's day, a theoretical basis for the existence of molecules made up of long chains of atoms had been formulated. That certain constellations of atoms ('radicals') were common to many organic compounds, and could perhaps be strung together to make larger molecules, was first envisaged

7

by one of the founding fathers of modern chemistry, the afore-mentioned Berzelius. It was also Berzelius who coined the term *polymer* (Greek for 'many parts') for compounds made up of identical proportions of atoms, but differing in size (molecular weight). The archetype was the family of compounds called olefines (on which a whole lot more later), with the formulae C_2H_4, C_3H_6, C_4H_8, and so on, or in general, C_nH_{2n}—though only the first and the third were known at the time. Berzelius had also made the critical observation that two quite different substances might yet have *identical* formulae. Such substances he called *isomers* (identical parts). The terms survive and will keep recurring in what follows.

In the latter half of the nineteenth century, new and more sensitive methods of molecular weight measurement evolved, and were soon being applied to colloids—among others, to rubber, to soluble, chemically modified starch, and to nitrocellulose, made by reacting cellulose, the fibrous material of wood and paper, with nitric acid. All returned the answer that the substances in solution were very large (with estimated molecular weights of anywhere up to about 100,000). This did not of course prove that they were not aggregates of small molecules, as the colloid chemists—an ever-growing band—still maintained, and the subject was bedevilled by the notion of 'partial valences'. These were the recently discovered, but then still ill-defined physical (as opposed to chemical) associations between atoms and molecules—something to which we shall return. They were commonly conflated with true chemical linkages (now called valence bonds, or *covalent bonds*).

The chemical nature of the known giant molecules was, at the same time, becoming clear, insofar at least as the building blocks from which they were constructed were being identified.

Figure 2 Jöns Jacob Berzelius (1779–1848), the great Swedish chemist, who coined the words protein and polymer.

And then some chemists found ways in which this process could be reversed, and simple chemicals be made to associate with each other to form long chains. Vinyl polymers and a kind of synthetic rubber had indeed been prepared as early as the mid-nineteenth century, but the sceptics had another way out of their dilemma: if molecules could combine head-to-tail with each other to start a chain, might the head not eat its tail and turn into a circular, and not too large molecule? This was a more palatable scheme to most chemists, for they were accustomed to viewing each chemical substance as unique, consisting, that is, of identical molecules. If the polymers were generated by successive addition of the reactive units (the *monomers*) in linear fashion, there would be no obvious way in which the chain length could be regulated. When all the monomers had reacted, one would be left with a collection of chains with a statistical distribution of lengths, some short, some long—a vision mightily offensive to orderly minds. Something of a landmark, though, was a lecture, delivered in 1863, and published three years later, by the influential French chemist Marcellin Berthelot,* proclaiming that there was nothing against the notion of indefinite covalent association of monomer units, and suggesting how it might be achieved.

It is hard now to conceive how the argument about whether macromolecules really exist could have raged until far into the twentieth century, especially since biochemists had long since put all doubts about the macromolecular character of proteins, carbohydrates, and even nucleic acids behind them. As late as 1929 a leading American colloid chemist could still trumpet the assertion that haemoglobin (and all other proteins) were aggregates of small molecules. What happened is that from about 1907 the waters were muddied by a rather disreputable group of chemists, led by Wolfgang Ostwald, the undistinguished son of a distinguished father, Wilhelm, one of the founders of physical chemistry. Wolfgang made up for his limited talent by an excess of ambition and proselytizing zeal. Colloids, he proclaimed, constituted a unique, hitherto unrecognized form of matter, distinct from all others, and not governed by the known physical laws. 'I see colloids everywhere,' he wrote. It was 'simply a fact' that colloids were the most universal and abundant of all the things we know. He went on to write three hefty textbooks on the subject. The message was taken seriously in some quarters. In Germany one of the most shifty, unappealing, and yet influential scientists of his time, Emil Abderhalden, a physiological chemist, made raucous

* Berthelot's contributions to chemistry were many and varied. He rejected on principle the idea that there might be a difference between natural and man-made substances, which would preclude the synthesis of the former in the laboratory, and went some way to proving his point. But Berthelot's interests extended beyond science. He had strong political convictions and in 1871 entered the Senate, having already served as president of the scientific defence committee throughout the siege of Paris by the Prussians, and later of the explosives committee. He subsequently became Minister of Education and for an unhappy period Foreign Minister.

pronouncements about the importance and ubiquity of colloids. A relentless self-publicist, like Ostwald, he invented a race- and sex-linked class of blood enzymes, traduced those who sought to prove them illusory, and finished up doing experiments on samples taken from captives in Auschwitz during World War II.

Another strong adherent of the colloid doctrine, of quite another calibre, was the top-ranking American physical chemist of the day, Arthur A. Noyes (known to his contemporaries as King Arthur), and many of the country's chemists followed his lead. Such endorsement enabled Ostwald to found a new learned society dedicated to the study of colloids and a journal to go with it. He spread the word in lectures around the world, and one of his American disciples even set up a profitable 'colloid healing' enterprise. Despite his ceaseless agitation Ostwald failed repeatedly to secure the professorial chair that he craved nor the directorship of an institute of the Kaiser-Wilhelm Society, the pan-German research organization, which transformed itself after World War II into the Max-Planck Society. Worldly success came to Ostwald, who had impeccable anti-Semitic credentials, only after he became active in the affairs of the Nazi party, following Hitler's rise to power in 1933.

Hermann Staudinger

Credit for forcing organic chemists to swallow the concept of giant molecules is generally allotted to another German, Hermann Staudinger, but there were other pioneers too. It was Staudinger, in fact, who coined the term 'macromolecule', which he preferred to polymers (the Latin versus the Greek derivation), setting off thereby a dispute, as heated as it was futile, about nomenclature. Today both are equally acceptable. Staudinger was an imperious man, temperamentally jealous of his scientific priority. His primacy in the field was ferociously defended after his death by his wife and collaborator, Magda Staudinger. He received the Nobel Prize for chemistry in 1953, and died at eighty-four in 1965. Staudinger was undoubtedly a man of principle. A pacifist during the Great War, he condemned especially the use of poison gases, and later he openly opposed the Nazis and deplored the renewed drive to war. In 1933, with the advent of the Third Reich, he was denounced, and the Rector of Freiburg University, the existentialist philosopher Martin Heidegger, an enthusiastic member of the Nazi party almost from the beginning, sought to have Staudinger dismissed. However, nothing came of this move, almost certainly because of Staudinger's relations with industry, for the development and manufacture of synthetic rubber became an urgent priority. So in the end the value to the state of the polymer chemist was recognized as surpassing that of the philosopher of being. Staudinger also had to contend with tireless machinations of Wolfgang Ostwald and another ardent Nazi and colloid chemist, Kurt Hess, who coveted his position. Relief came in 1943 with the unlamented demise of Ostwald.

One of Staudinger's arguments for the existence of polymers was that rubber, for example, retained its high molecular weight when measured in a wide range of solvents, some of which, surely, would have been expected to break the 'partial valences' if these were what held the colloid together. But it was the synthesis of fibrous polymers that finally defeated what opposition remained.

It is now time to take a grip on the fundamentals.

2

THE BASICS: A LITTLE CHEMISTRY

The carbon world

The fundamentals are simple enough. Understanding the nature of polymers, whether proteins, carbohydrates, DNA, or synthetic fibres and moulded plastics, requires first of all a modest grasp of the rules governing the architecture of organic molecules. Chemical formulae first, without regard to geometry: water, H_2O, has the structural formula $H-O-H$, where the dashes represent valence (or covalent) bonds. Oxygen, with its valence of 2, can form attachments to two other atoms, and hydrogen to only one. Gaseous hydrogen is H_2, thus $H-H$, and gaseous oxygen is O_2, thus $O=O$, where the two dashes represent a *double bond*, with both valences engaged. Carbon has a valence of 4, so the simplest carbon compounds can be written CH_4, or

$$
\begin{array}{c}
H \\
| \\
H - C - H \\
| \\
H
\end{array}
$$

for methane, and $O=C=O$ for carbon dioxide. Nitrogen has a valence of 3, and its simplest compound is ammonia, NH_3, or

$$
\begin{array}{c}
H \quad\; H \\
\backslash\;/ \\
N \\
| \\
H
\end{array}
$$

These are simplified rules: for most elements (though not carbon or oxygen) more than one valence state is permitted, and the situation is in reality more complex generally, but the simple scheme will serve. So let us move on to more interesting compounds.

Carbon, as we have seen, has the capacity to form long chains. With two carbons and hydrogen we get ethane,

$$
\begin{array}{c}
H \quad H \\
| \quad\; | \\
H - C - C - H \\
| \quad\; | \\
H \quad H
\end{array}
$$

with three carbons propane, with four butane, and so on. This *homologous series* is a family of compounds called paraffins, or *alkanes*. With a very large number of carbons we arrive at polyethylene, of which much more shortly:

$$
\begin{array}{c}
H \quad H \quad H \quad H \qquad H \\
| \quad\; | \quad\; | \quad\; | \qquad | \\
H - C - C - C - C \cdots - C - H \\
| \quad\; | \quad\; | \quad\; | \qquad | \\
H \quad H \quad H \quad H \qquad H
\end{array}
$$

But it is also now obvious that, once we get beyond three carbon atoms, they can be linked in different ways. So butane, the fourth member of the alkane series, can take two forms, one linear, one branched:

$$
\begin{array}{cccc}
 & H & H & H & H \\
 & | & | & | & | \\
H- & C- & C- & C- & C-H \\
 & | & | & | & | \\
 & H & H & H & H \\
\end{array}
\qquad
\begin{array}{ccc}
 & H & \\
 & | & \\
 & H-C-H & \\
 & H \quad | \quad H & \\
 & | \quad | \quad | & \\
H- & C-C-C & -H \\
 & | \quad | \quad | & \\
 & H \quad H \quad H & \\
\end{array}
$$

They have the same formula, C_4H_{10}, but they are different substances with significantly different properties. The first is *n*-butane, the second is *iso*butane, and they are *isomers* of one another. The next higher alkane, C_5H_{12}, pentane, has three isomers, and so on.

Returning now to polyethylene, this polymer (which is known commercially as polythene) takes its name from the compound from which it is synthesized, namely ethylene:

$$
\begin{array}{ccc}
H & & H \\
\diagdown & & \diagup \\
 & C = C & \\
\diagup & & \diagdown \\
H & & H \\
\end{array}
$$

This is the first of the family of olefins, or *alkenes*. Because the carbon atoms here carry fewer than their maximum possible number of hydrogens, this type of compound is termed *unsaturated*, whereas the alkanes are *saturated* compounds. (The terms saturated and unsaturated have entered common usage in relation to the 'good' and 'bad' dietary fats: the 'good' ones contain carbon–carbon double bonds.) An important difference between the properties of single and double bonds is that the atoms or groups of atoms at either end of a single bond can rotate freely about that bond. A double bond does not permit such unrestricted rotation: imagine two balls with a hole in each, into which a rod is fitted. These balls can be rotated without impediment. But if the balls are linked by two parallel rods, the system is rigid.

A vast and important class of unsaturated compounds are the *aromatic* hydrocarbons—the name is historical, not literal—of which benzene is the archetype. They are made up of rings. The formula for benzene is written

and when, for instance, a hydroxyl group, −OH, is substituted for one of the hydrogen atoms:

this becomes phenol, a compound with quite different properties. Rings may be fused, as in the simplest case, which is naphthalene (moth balls):

and the rings may also contain other atoms, commonly nitrogen, as in the bases of DNA

pyrimidine

purine

which are respectively the parent molecules of the famous C, T and A, G, cytosine and thymine, and adenine and guanine:

cytosine

thymine

17

adenine

guanine

Left hand, right hand

There are two further features of both natural and man-made compounds to take into account. The first is that molecules are not after all flat, as they appear on paper, but exist in three dimensions, and have a geometrical form. This, at the time revolutionary, notion, which now seems so obvious, was advanced independently by Joseph Achille Le Bel in France and Jacobus Henricus van't Hoff in Holland around 1878. It divided the world of chemistry into sceptics and adherents. Among the sceptics were many of the most influential German organic chemists (and in that science Germany led the world). The most vociferous was Hermann Kolbe, a notoriously intolerant and evil-tempered man, who denounced the proponents of the theory in incandescent terms:

[t]wo practically unknown chemists, one from a veterinary school, the other from an agricultural institute, judge the deepest problems of chemistry, which will probably never be answered. They judge these highly important problems, particularly the question of spatial orientation of the atoms, with a cocksureness and insolence that can only astonish a true student of the natural sciences.

Van't Hoff was in time recognized as a scientific colossus, and indeed became in 1901 the first to win the Nobel Prize for chemistry.

This was the question that van't Hoff and Le Bel asked themselves: if the dispositions of the valence bonds of an atom have a defined geometry, how, for instance, are the four hydrogens which are bound to the carbon atom in methane arranged in space? Their inferred answer was *symmetrically*, based on the reasonable assumption of perfect equivalence of all the hydrogens. Now, disregarding a flat arrangement with the four hydrogens at corners of a square, there is only one fully symmetrical way to surround an object by four identical other objects: they must be placed at the corners of a *tetrahedron*:

Out of these cogitations (in time fully vindicated by experiment) grew the subject of *stereochemistry*, which concerns itself with the arrangement of the atoms of a molecule in space. (The word derives from the Greek *stereos*, meaning solid, thus implying three-dimensional.) Stereochemistry transcends mere formulae, for it tells us how molecules look, and not merely how their constituent atoms are linked, and it soon provided the answer to one of the great conundrums of chemistry, physics, and even biology.

Suppose then that, instead of just four hydrogens, four different atoms or groups are bound to that carbon, call them *a*, *b*, *c*, and *d*. There are two ways of arranging these four groups around the central carbon atom:

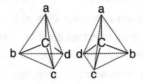

These two objects are not identical; no matter which way you turn them, they cannot be superimposed on each other any more than a left hand can on a right. They are in fact mirror images. They are *enantiomers* (from the Greek for opposite), also called *stereoisomers*. The amino acids from which all proteins are constructed have this property, and their two forms are designated L and D, deriving from the Latin *laevus* (left) and *dexter* (right), based on the direction in which they rotate the plane of polarized light.

Polarized light needs a little explanation. The phenomenon has been known since at least the mid-seventeenth century, and probably a lot longer. Light waves oscillate uniformly in all directions, but when they pass through an intrinsically asymmetric medium, such as crystals of the mineral Iceland spar in which the effect was probably first seen, the waves exit oscillating only in one plane; they have, so to speak, been flattened. The polarized light ray can then pass unimpeded through an Iceland spar crystal with its axis oriented in the same direction as that of the first crystal; however, one with its axis at right angle to the first will be opaque to the polarized light. This effect is used in sunglasses, for light from a bright sky is polarized. When sunlight is reflected from, say, the surface of water, the plane of polarization is to a large extent horizontal. So a Polaroid film (a modern equivalent of the Iceland spar crystal) with a vertical orientation will screen out the glare. Bees are one of the creatures that can sense a plane of polarized light, and use it to navigate by. In about 1815, the French physicist Jean-Baptiste Biot discovered that solutions of many natural substances (though none of those synthesized in the laboratory, but see later) rotated the plane of polarized light passing through them, some in one direction, some in the other. The polarimeter, the instrument devised by Biot to measure the rotation, soon became an important tool for organic and physiological chemists.

To resume then, the principle of stereoisomerism was discovered by Louis Pasteur, then aged twenty-six, in one of the prettiest experiments in the history of science. In 1848, engaged in the study of fermentation processes, he prepared from tartar a by-product of wine manufacture, crystals of a salt of tartaric acid, watery solutions of which were known to be *laevo*-rotatory. When he inspected the preparation he observed something that had escaped his predecessors—the crystals were all asymmetric in the same sense, and not a mixture of mirror-image objects, as illustrated. *

Figure 3 Left-handed and right-handed quartz crystals: mirror images of each other.

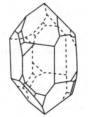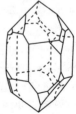

* These crystals have a shape known to geometricians as hemihedral, first observed in the mineral quartz. The two forms of quartz rotate the plane of polarization of polarized light in opposite directions. This relation between crystal form and optical activity was perceived by the astronomer Sir John Herschel in 1820.

21

Could the asymmetry of the crystals then somehow reflect the direction of rotation of polarized light? Pasteur's intuition was correct. Then came his second remarkable insight. Racemic acid, another product of fermentation (named from the Latin, *racemus*, a grape), was chemically identical to tartaric acid but had no effect on polarized light; might it then be a mixture of left- and right-handed enantiomers? And so it proved, for the crystals of its salt contained equal numbers of the two mirror-image forms. With tweezers and a magnifying glass Pasteur was able to sort some of the crystals into two small piles of the opposing types. The hemihedral faces, he wrote, were 'sometimes inclined to the left and sometimes to the right'. From each he prepared the soluble salt and lo, they rotated the plane of polarized light in opposite directions.

This was a stunning discovery, so striking that the French Academy of Sciences demanded visual and attested proof, and they deputed an Academician, a veteran physicist and leading expert on the properties of light, to watch Pasteur perform his experiment: it was Jean-Baptiste Biot, who discharged his task with diligence. He brought his own preparation of racemic acid and the reagents needed to prepare crystals to a laboratory in the Collège de France, where Pasteur awaited him. He watched while Pasteur prepared the solution of the salt, which was then left in a locked room to evaporate and crystallize. When Biot saw that the crystals had formed, Pasteur was summoned to perform his trick with magnifying glass and tweezers. From the two piles of crystals which resulted Biot prepared solutions and measured the angles of the plane of polarization. The result was recognized at once as one of the most spectacular coups in the history of science. Biot found the result so beautiful that he was choked with emotion. Pasteur recalled the old man's words:'Mon enfant, j'ai tant aimé les sciences dans ma vie que cela me fait battre le cœur.' ['My child, I have so loved science throughout my life that this makes my heart thump.']

Natural molecules are nearly all asymmetrical in this sense: all proteins are made up of L-amino acids, and nucleic acids of D-nucleotides. As we shall see, biological polymers such as these generally are, or contain, long asymmetric spiral (helical) structures. Clearly for such a spiral to assume a left- or a right-handed screw sense, its constituent parts must be asymmetric to start with; otherwise neither one or the other would be preferred. So a protein chain, which consists of L-amino acids, will tend to form a right-handed helix, whereas, if it is made of D-amino acids, it will assume the opposite (mirror-image) symmetry. Substances synthesized from simple starting compounds by organic chemists are by contrast optically inactive. If such products are intrinsically asymmetric, they will contain equal proportions of D- and L-enantiomers. This is called a *racemic mixture*, reminding us of Pasteur's discovery. To separate the two components tweezers and magnifying glasses are no longer used; less laborious methods are available, involving chemical reaction with, or attachment to, molecules or matrices which are themselves asymmetric and pick out one or other enantiomer. Because living matter is asymmetric, metabolites and drugs must also have the correct symmetry to exercise their function. The importance of this rule is illustrated by the distressing thalidomide episode; the drug was marketed in the form of a racemic mixture. Only one of the two enantiomers was responsible for causing birth defects when ingested (as a tranquillizer by pregnant women), whereas the other, had it been separated from the mixture, would have been a harmless and effective drug.

23

How enantiomeric forms of the molecules of life emerged when the Earth cooled has been a matter of inconclusive debate for many decades. It was shown some fifty years ago that when a mixture of the gases of which the early atmosphere largely consisted (methane, ammonia, water vapour, and a few others) was subjected in the laboratory to electric discharges, to simulate conditions that might have occurred during thunderstorms, organic molecules, including amino acids, could be formed. But there is no obvious reason why these should appear in any form other than a racemic mixture. Yet it was found that the Murchison meteorite (which landed in Victoria, Australia, in late September 1969) contains some organic materials, including slightly more of an L-amino acid than its D-enantiomer. A favoured conjecture is that the polarization of light in the solar system by magnetic fields may have been responsible for preferential formation of the former. But this is only one of many theories, none of them so far substantiated. (One notion is that a primitive life form originated elsewhere in the cosmos and was brought to the Earth on a meteorite.)

The hydrogen bond

The second inescapable feature of natural (and most other) molecules is that the partial valences alluded to earlier as a historical aberration do exist. That is to say, there are well-defined physical forces between atoms within or between molecules, far weaker than covalent bonds, but still important. Predominant among these is the *hydrogen bond*, which can form between a hydrogen atom and an oxygen or nitrogen atom. It is the hydrogen bond that determines the unique properties of water. Without it water would be a gas at the temperature at which we thrive, and there would be no life. So water molecules in the liquid or frozen state are not simply H—O—H, but form transient, ever-changing clusters, such as this, in which dotted lines denote the hydrogen bonds:

You are now equipped for the first part of our journey round the world of giant molecules.

3

PROTEINS: FROM SKIN AND BONE TO ENZYMES

We will not be concerned here with biochemistry—with what proteins do in the metabolism—but rather with their structure. In the days of Mulder and Liebig the prevailing wisdom was that the proteins of the body, such as albumin and haemoglobin, were ingested, whether from steak or from spinach, ready made, requiring at most a little touching up before being put to use. This is the opposite of the truth, for proteins in our food, the principal suppliers of the body's nitrogen, are broken down by a series of metabolic reactions. The products of this degradation are then used to make new protein and other molecules of life. But animal and plant proteins are constructed along the same lines from the same twenty L-amino acids.

So far all the polymer chains we have considered have had back-bones consisting of only carbon atoms. But there are some—and they include the most interesting biological polymers—in which the carbons are interspersed with atoms of other elements. First among these is the *polypeptide* chain, on which all proteins are based. Amino acids, the fundamental building blocks of proteins, have the formula

or in shorter form: $NH_2.CHR.COOH$. The group

written NH_2, is called an amino group, and the group

or COOH, is a carboxyl group, while R is the *side chain*. But in reality the first two types of group are not quite what they seem, for they are respectively *basic* and *acidic*, and both carry an electrical charge in neutral watery solution, in which biological substances most often find themselves. To explain this requires a digression into the nature of acids and bases.

Vitriol and vinegar

Strong acids, notably sulphuric acid ('oil of vitriol'), hydrochloric acid ('spirits of salt'), and nitric acid (*aqua fortis*, or 'strong water'), were known to the alchemists, but it was not until the eighteenth century that the first dim light was shed on what united them. Antoine Lavoisier, who has a fair claim to be regarded as founder of quantitative chemistry, and who met his end on the scaffold during the French Revolution, thought that the essence of all acids was oxygen. The word *oxygène* was his, meaning bringer of sharpness. Classical scholars would have viewed this as a barbarous coinage (two centuries later one critic observed about television that no good would come of an invention the name of which was half in Latin and half in Greek), and moreover Lavoisier was wrong. Yet the name stuck, and in German oxygen became, and still is, *Sauerstoff*—sour substance. Another century passed before the nature of acidity was (more or less) fully explained by another of the towering figures in the history of chemistry, the Swede, Svante Arrhenius. The truth was nearly all contained in his doctoral thesis, presented to Uppsala University, and for which he was awarded only a fourth-class degree, carrying the niggardly approbation of *non sine laude approbatur*, 'approved, not without praise'. It set out the essence of electrolyte theory, one of the foundations of physical chemistry; although generally rejected at the time, it later won him the Nobel Prize in 1903.

The nature of atoms was not then known. It became clear only a decade later, from the work of Ernest Rutherford—the man who emerged from a sheep farm in New Zealand to rise to his apotheosis as Lord Rutherford of Nelson, Nobel Laureate and President of the Royal Society—that atoms are mostly empty space. They consist of a positively charged nucleus, in which nearly the entire weight of the atom resides, surrounded by negatively charged electrons viewed in the original model as orbiting round the nucleus like planets around a sun—the image appears in innumerable logos. The atom is electrically neutral, because the number of electrons balances the number of protons in the nucleus, with their equal and opposite charge. (The nucleus, except in the case of the lightest element, hydrogen, also contains neutrons, which are uncharged but equivalent in weight to the protons.) Arrhenius posited that salts, which result from the association of an acid with a base, are dissociated into positively and negatively charged *ions*, when dissolved in water. So common salt, sodium chloride, NaCl (which can be made by mixing hydrochloric acid, HCl, with sodium hydroxide, NaOH) is not really NaCl, but is made up of equal parts of sodium ions and chloride ions, written Na^+ and Cl^-. In other words the sodium atom loses an electron, which the chlorine atom gains.

This holds even in the common salt crystal, the structure of which was solved in 1913, the first triumph of the new science of X-ray crystallography, for which the Braggs, father and son, received the Nobel Prize. (Lawrence Bragg, the son, heard of the award while serving on the Western Front at the height of the Great War.) In the sodium chloride crystal each sodium atom is surrounded equidistantly by six chlorines, and vice versa, so there is no Na—Cl bond, only Na^+ and Cl^- ions, held in place by electrostatic attraction. This discovery caused distress to many chemists, amounting to panic, and Lawrence Bragg related that one of their number even pleaded with him to look again at his crystal and try to move the sodium at least a little bit closer to one or other chlorine and further from all the rest.

This dissociation of salts into the constituent ions is respon- sible for the capacity of their solutions in water to conduct electricity. They are in fact *electrolytes*. In an electric field the ions move towards opposing electrodes, positive ions to the cathode, negative to the anode; hence their name, from the Greek *ion*, a wanderer. It is the salts (mainly of calcium) dissolved in tap water that make it inadvisable to think of warming your bathwater by plunging an electric fire into it (a practice in the early days of mains electricity that caused many a fatality).

So how does all this relate to acids and bases? What makes a solution acidic—sour to the taste when dilute, like vinegar or lemon juice—is the presence of hydrogen ions. The hydrogen atom has a single proton for its nucleus and an electron outside. The hydrogen ion is therefore merely a proton. The acidity of a solution is measured by the concentration of protons, normally expressed on a logarithmic scale (multiples of 10), called pH, which was invented by Arrhenius. That is to say, a pH difference of 1 corresponds to a tenfold difference in proton concentration. A neutral solution is characterized by a pH of 7. A lower pH denotes the acidic, a higher the alkaline state. Now, according to Arrhenius, an acid is a molecule (or group in a molecule) that can give up a proton, H^+. A strong acid releases its proton easily, a weak acid more reluctantly. A base, on the other hand, is a molecule or group that can release a hydroxyl ion, OH^-. A very strong base is referred to as an alkali: the most familiar example is sodium hydroxide ('caustic soda'), NaOH. An acid will neutralize a base, causing H^+ and OH^- to combine to form H_2O. But, the definitions of acid and base were later broadened; an acid is now defined as a molecule (or group) that can lose a proton or take up a hydroxyl ion, while a base can lose a hydroxyl ion or take up a proton. With all this in mind we can now return to the amino acids.

The nature of amino acids

Amino acids are both acids and bases, and they exist in neutral solutions in the doubly ionized form (called a zwitterion, from the German for a hybrid). So the amino and carboxyl groups are properly written respectively $-NH_3^+$ and $-COO^-$. The amino acids, then, have the structure

$$H_3N^+ - \underset{\underset{H}{|}}{\overset{\overset{R}{|}}{C}} - COO^-$$

But when linked together to form a polypeptide chain, it is only (disregarding, for the moment, the side chains) the chain ends that retain a charge, an amino group at one end, called the N-terminus, and a carboxyl group at the other, the C-terminus. In between are the uncharged peptide groups. The structure looks like this:

$$H_3N^+ - \underset{\underset{H}{|}}{\overset{\overset{R}{|}}{C}} - \underset{\overset{\|}{O}}{\overset{}{C}} - \underset{\underset{H}{|}}{\overset{\overset{H}{|}}{N}} - \underset{\underset{H}{|}}{\overset{\overset{R'}{|}}{C}} - \underset{\overset{\|}{O}}{\overset{}{C}} - \underset{\underset{H}{|}}{\overset{\overset{H}{|}}{N}} - \underset{\underset{H}{|}}{\overset{\overset{R''}{|}}{C}} - \underset{\overset{\|}{O}}{\overset{}{C}} - \underset{\underset{H}{|}}{\overset{\overset{H}{|}}{N}} - \underset{\underset{H}{|}}{\overset{\overset{R'''}{|}}{C}} \cdots - \underset{\overset{\|}{O}}{\overset{}{C}} {\overset{\diagup O^-}{}}$$

where R, R', R″, etc., are different side chains. A more economical way of representing this polymer is in terms of the repeating unit:

$$\left(\underset{\underset{H}{|}}{\overset{\overset{H}{|}}{N}} - \underset{\underset{H}{|}}{\overset{\overset{R}{|}}{C}} - \underset{\overset{\|}{O}}{\overset{}{C}} \right)_n$$

Here n, the number of amino acids that go into making up the protein chain, can be very large. It is the side chains that distinguish the twenty amino acids from each other. In the simplest of them, glycine, the side chain consists just of hydrogen, H, whereas the other nineteen have carbon-containing groups, three of them in the form of rings, and five carrying charges.

One further feature of the amino acids is that all of them, except for glycine, possess a central carbon atom with four different groups attached to it: an amino group, a carboxyl group, a side chain, and a hydrogen. They are, then, all compounds of the type depicted earlier (p. 20), and this is the reason they can exist in either of two mirror-image versions, L or D. It bears repeating that we live in a left-handed world: all proteins are made up solely of L-amino acids. A mirror-image protein would not in general be able to fulfil its functions, which require it most often to interact with existing left-handed proteins, and it would be of little use in a cell.

Each of the many proteins in our body is characterized by a unique sequence, from which its properties derive. Unlike long polymers synthesized in the laboratory, the length of the chain of each protein is fixed. A typical small protein (the enzyme ribonu-clease) has a chain length of 124 amino acids; the largest proteins run to thousands. We will have more to say about the disposition of these building blocks, commonly referred to, since they are no longer isolated amino acids, as amino acid *residues*. With twenty amino acids to choose from, there is an effectively infinite number of possible sequences in which they can be arranged (the amino acid sequence, also called the *primary structure*). The amino acid sequence of each protein is determined, as we shall see, by a corresponding sequence of chemical units (the nucleotides) in the genetic material, the DNA. The sequence of nucleotides that specifies a particular protein is nothing less than the gene for that protein. From time to time a mutation—the conversion of one nucleotide to another by a metabolic accident, or by the action of an intruder, such as a carcinogenic chemical or radiation— afflicts the gene, and is most commonly reflected, for good or ill, in the exchange of one amino acid in the sequence for a different one.*

What does a protein look like?

If one were to make a polypeptide chain with its twenty kinds of amino acids in a random sequence, and if it were soluble in water (something that depends on several factors, but most of all on the proportions of different amino acids in the chain), that chain would look as floppy and disorganized as a piece of string or vermicelli. It would be what is loosely referred to as a *random coil*, in a constant state of motion when in solution, under the bombardment of the surrounding water molecules, all hurtling around in restless agitation. Such a protein would generally be of little use to a living organism, but 3 billion years of mutational trial and error have changed all that. Evolution has provided us with proteins that are not random coils, but have tightly constrained structures. This comes about because the units in a polypeptide chain, when arranged in an appropriate sequence, will interact with each other. The interaction, depending on the amino acid sequence, can take the form of a regular, ordered structure, or merely of an intimate cluster.

---

* This is a gross simplification, which might offend geneticists. Some mutations may also cause the sequence to be prematurely terminated, so that only a truncated protein is produced, or the entire sequence after the site of the mutation may be altered into something quite different and useless. Mutations are the agents of evolution. Of course, many do not have any effect at all.

The regular patterns which a polypeptide chain can assume were first divined by the American chemist Linus Pauling in 1951. Because of the nature of the covalent bonds between successive amino acid residues that make up the backbone of the chain, only certain angles of twist are possible. In addition it is highly favourable for the chain to assume a shape which allows hydrogen bonds to form between residues. The only groups in the polypeptide backbone capable of forming hydrogen bonds are $-C=O$ and $-N-H$, which can associate like this: $-C=O \ldots H-N-$, with all bonds aligned. As Pauling tells it, inspiration came to him while he was on sabbatical leave, enduring the cold and damp of postwar Oxford, and in bed with a cold. Bored with his detective novel, he fell to thinking about protein structure, and summoned his wife to bring him paper, pencil, ruler, and scissors. Because the lengths of the covalent bonds and the angles between them were known, or could be inferred from the structures of simple compounds related to the peptide bond, Pauling was able to cut out a relatively accurate representation of a protein backbone. He then twisted his strip of paper to bring the $N-H$ and $O=C$ groups into apposition, so that a hydrogen bond might form between them. This was made easier by the known length of a hydrogen bond and because it requires the $N-H$ and $O=C$ to be in line. By this rudimentary means, backed up later of course by accurate model building with metal rods, Pauling discovered three geometries giving minimal distortion of all the bond lengths and directions. The first of these instantly became famous (to the great chagrin of Lawrence Bragg and his colleagues in Cambridge, who had got it wrong) as the right-handed α-helix (pronounced alpha-helix). Here the dashed lines represent hydrogen bonds:

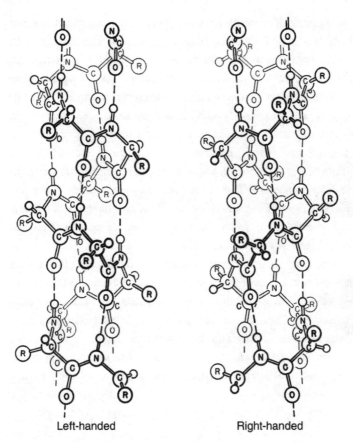

Left-handed   Right-handed

Figure 4 The α-helix.

It can be seen from the picture that each residue makes a hydrogen bond with the fourth one along in the chain, but the geometry is such that there are eighteen residues for every five turns of the helix, in other words 3.6 residues per turn. (This was where the Cambridge group went wrong, for they assumed that a repeating unit of any structure would contain an integral number

35

of residues.) The asymmetry here is critical, for a chain made of L-amino acids forms a helix with a right-handed screw sense. If natural proteins were composed of D-amino acids, then symmetry dictates that the helix would be left-handed. (Other types of helix with different numbers of amino acids per turn can theoretically be made, but are less favourable and are only rarely seen in life. The important exceptions are certain proteins of unusual composition, notably collagen, which we will discuss later.) The fact is, though, that not all amino acid sequences are compatible with the α-helical structure. This is mainly because some side chains are large and would crowd each other out if packed into a helix, or if many of them carry positive or negative charges, their proximity in the helix would lead to strong repulsion.

The other structures that Pauling discovered by model-building are the β-structures (pronounced beta), also known as the pleated sheet, from its appearance when many lengths of chain associate in parallel. They are formed by side-to-side association of almost fully extended chains, and the reason that two forms are possible is that adjacent chains may run in parallel or antiparallel (i.e. the same or opposite) directions, as shown here:

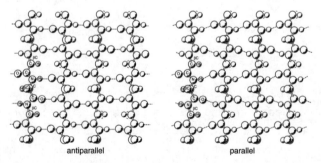

antiparallel   parallel

Figure 5 The β (the plated sheet) structures.

A β-structure, necessarily antiparallel, may also be formed within a single chain, when it doubles on itself to form a hairpin; or, if the chain is long enough, two remote parts of the sequence may also be able to unite in a parallel β-structure. The α-helix and the β-sheet (to which one may add the special configuration of the chain that allows it to make the hairpin bend, and very rare bits of freak helices seen in some proteins) define the *secondary structure* of the protein. The key to protein structure was the recognition that it is uniquely determined by the amino acid sequence—the primary structure—alone. Some sequences define fibrous, and others (the majority) globular proteins, and we will now consider their characteristics and functions.

Fibrous proteins—fingernails, leather, and the ingenious spider

The first known structure of a fibrous protein was that of keratin. It comes in two forms, α- and β-keratin, and their structures are what their designations imply. α-Keratin is the major protein of hair, nails, feathers, and skin, and is entirely α-helical, but it also has a higher order structure, with three strands of α-helix twisted around each other like strands of spaghetti. This results in a filament of increased stiffness and tensile strength. Leather is practically all α-keratin. β-Keratin, on the other hand, is largely made up of antiparallel β-pleated sheets. It is the principal protein of birds' beaks and claws and of silk.

The structure of silk has been thoroughly studied and, as natural fibres go, is rather simple. In the first place, the composition is strongly biased towards amino acids with the smallest side chains: about 60% are glycine, the smallest of all, in which a mere hydrogen atom serves as side chain, and alanine, the next small-est, in which the side chain is a methyl group, $-CH_3$. These two alternate in sets of six: ser-gly-ala-gly-ala-gly. The first member of this set is serine, which has a slightly larger side chain, $-CH_2OH$, or more explicitly:

$$\begin{array}{c} \text{H} \\ | \\ \text{O} \\ | \\ \text{H}-\text{C}-\text{H} \\ | \\ -\text{N}-\text{C}-\text{C}- \\ \hspace{0.3em}| \hspace{0.8em} | \hspace{1.2em} \| \\ \hspace{0.3em}\text{H} \hspace{0.7em}\text{H} \hspace{1em}\text{O} \end{array}$$

The chain contains ten such segments in succession: $(-\text{ser-gly-ala-gly-ala-gly-})_{10}$, making up sixty amino acids. These sequences of ten blocks are interspersed with segments of quite different character, made up of amino acids with varied, in general much larger, side chains. Now the runs of amino acids with small side chains allow the polypeptide chains to pack comfortably into β-sheets, whereas the intervening sequences preclude formation of ordered structure, and so nestle between the β-sheets as random coils. In silk the polypeptide chains that make up the β-sheets are aligned in the direction of the spun fibres, and because of the serried hydrogen bonds that link them, they confer great tensile strength on the material. But the intervening random coils are easily stretched, and so permit a degree of elastic extension when the fibres are subjected to a sharp tug. This makes them highly suitable for use in sewing thread for instance.

The silk we are familiar with originated in China during the third millennium BC, but it was to be three millennia before

Figure 6 Schematic picture of silk protein, showing blocks of β-structure (pleated sheet) interspersed with segments of flexible random coils.

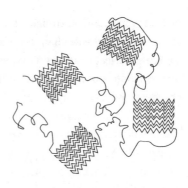

missionaries illicitly brought cocoons of the silkworm, *Bombyx mori*, to Europe. The fact, though, is that many insects and spiders make silk, and in recent years there has been much interest in spider silk. The notion of spider farming for silk was first broached in the early eighteenth century; indeed, in 1709 a French weaver made a pair of gloves and one of stockings from spiders' silk and presented them to the Sun King, Louis XIV. But commercial manufacture was never really on the cards because of the huge number of spiders needed to produce even a small amount of silk. Today, though, the methods of genetic engineering allow the synthesis of alien proteins in bacteria, in plants, and in the milk of farm animals, so the vision of spider silk farms presents itself anew. So far, although more than one biotech company has raised money for such a venture, nothing useful has emerged. One problem is that the spider makes the protein as a very concentrated solution, and most often the end-product is a composite, involving two materials, one of which coats the other. This is beyond the powers of a bacterium.

The spider is equipped with a system of pumps and valves that allow the viscous, freshly synthesized protein solution to be squirted through the orifice of an organ called a spinneret. As it emerges the filaments of protein align in the direction of flow, making in effect a liquid crystal, which, because it is so thin, dries and solidifies on the way out. The spider has some control over the fibre properties, for it can adjust the pumping pressure and the mixing proportions where there is more than one component, but for different elements of its web it will generate different fibre types. Most fibres of the garden spider are about 1 μm (commonly referred to as a micron) across, that is, about one-thousandth of a millimetre. Some other species of spider make much finer threads, whereas some monsters found in Asia generate huge, dense webs of thicker filament, in which animals or men can become entangled and find some difficulty in escaping. The fact is spider silk is both astonishingly strong and retentive. It is often asserted that it has greater tensile strength than mild steel, but this must be qualified. Judged in terms of strength per unit cross-sectional area, it is a lot weaker, but then its density is little more than one-tenth that of steel, so weight-for-weight it is indeed stronger.

From the viewpoint of polymer chemistry, the ingenuity of the spider far exceeds that of the chemist. Spider silk, like that of the silkworm, is derived from a form of keratin. The filament protein is made up of segments of hydrophobic (literally water-hating) amino acids with small side chains, which can fit into β-pleated sheets, interspersed with tracts of hydrophilic (water-loving) amino acids in the randomly coiled state. (We will come to what makes a group hydrophobic or hydrophilic later.) Pleated sheets are embedded in the spaghetti-like coiled loops of chain, resembling (as will become plain in due course) rubber. So when pulled, the coiled segments can stretch out and the fibre will elongate. It will not easily break, because fracture of a structure (like, for instance, a girder of a bridge) follows only when a crack occurs. The crack induces a local structural disturbance, and it is there that resistance to mechanical stress is catastrophically impaired. In composite materials, about which more presently, cracks are far more likely to be averted or sealed. Its composite character gives the spider thread (and that of the silkworm) its tensile strength. But the demands made on the web's capture threads transcend mere strength. Certainly the thread must survive the impact of a high-velocity insect, but if it does not deform, the missile will merely ricochet. And if, on the other hand, it recoils as sharply as rubber it will propel the insect back out as from a trampoline. So the rate and extent of recoil must be finely tuned if the spider is not to go hungry.

The capture thread of some common spiders has an additional safeguard against the insect's escape: it is garnished with globules of liquid glue. This substance is squeezed out of a nozzle, close to where the silk protein emerges, and coats the drying thread, but the restricted wettability of the surface (like that of Teflon) causes the liquid film to break up into droplets. These can be seen, strung like beads along the thread at more or less equal intervals. The liquid adhesive is made up of a glycoprotein—a protein with chains of sugars chemically attached to certain of the amino acid side chains—and some smaller molecules that take up and retain large amounts of water. The properties of the finished capture thread, then, are truly remarkable: it can contract to a fraction of its resting length and stretch to at least four times that length before it breaks, so is capable of a tenfold or greater length change. When caused to stretch, it acts like rubber up to a point at which it, quite abruptly, becomes rigid. When the deforming force falls off, that is when the insect has been brought to a halt, the threads relax, but gently, not like a rubber band. Nor do the threads droop while this occurs, and the relaxation in fact appears to be regulated by the local coalescence through surface tension of the glue droplets.

The strongest fibre of most spiders is the dragline, by which the creature suspends itself and which also forms the outer frame thread of its web. This is constructed around a core consisting of two proteins, which differ in composition from those of the other threads and also those of the spider's egg sacs with their silk covering. One of the remarkable properties of the dragline is its resistance to a twisting force—it barely twists when the spider descends, and if a twist is imposed on it, it oscillates by only a few degrees about its new position. (As the engineers would put it, the thread exhibits highly damped oscillation.) Not only that, it will then return gradually to its original orientation, and unlike a metal wire, is scarcely subject to fatigue, for it will tolerate repeated applied twists. It thus bears little relation in its behaviour to any man-made materials.

This is not the totality of the extraordinary versatility of spider silks. Other species of spider produce dry fibres with a twentieth of the diameter of the common garden spider. To achieve this they fluff up the silk by passing it through fine combs on their back legs. These thin filaments also have a complex structure, but they break when an insect enters the mesh, and catch it by its legs or hairs. The black widow spider, by contrast, anchors its web to the ground with glue-coated, fully stretched threads. The impact of an insect causes them to break, whereupon they at once contract like a broken rubber band and catapult the prey into the web. Little wonder then that biotechnologists are eager to mimic these properties with synthetically produced spider proteins.

The body's proteins—rigidity, elasticity, and the jumping flea

All of us are held together by fibrous proteins. A quarter of our body protein is *collagen*, possibly the most abundant protein on earth. There are actually at least twenty collagens, all very similar in structure, but differently distributed in the animal body. Like silk, collagen is rich in glycine, which occupies every third position in the sequence, and the other abundant amino acid is proline. This amino acid is unique in that its side chain makes a loop, attached at its other end to the nitrogen atom of the polypeptide backbone; thus:

$$
\begin{array}{c}
{}^{\diagup}CH_2{}^{\diagdown} \\
CH_2 \qquad CH_2 \\
{}_{\diagdown} \qquad {}_{\diagup} \\
NH-CH-COOH
\end{array}
$$

The configuration of the backbone is consequently grossly restricted around each proline, and with so many prolines in the chain, collagen is forced to adopt a unique helical conformation of its own. The chain contains 1,050 amino acids, and its sequence consists of the repeated triplet pattern, glycine-proline-X, where X can be any of the other amino acids. This chain forms a left-handed helix, and three of them fit snugly together to make a right-handed rope, which is very stiff—much more so than an α-helix. Many of these ropes, which are called *fibrils*, then pack

43

together to form thick cables, commonly several μm (micrometres) in diameter, and visible in the light microscope. There are mutations, which put alien amino acids into positions properly occupied in the sequence by glycine or proline; the formation of fibrils is thereby impaired and disorders of the skin, of tendons, ligaments, or bones are the result. (The most common is Marfan's syndrome, which gives rise to abnormally extensible joints, said to have accounted for both the Mephistophelian appearance and the preternatural virtuosity on the violin of Niccolo Paganini. Marfan's syndrome is associated with elongated facial features: Abraham Lincoln is one who was supposed to have suffered from the condition.)

The collagen fibrils, and the thick fibres even more so, are very rigid and resistant to tensile stresses. This property makes them a highly suitable material for tendons, for arteries, which must endure the powerful hydraulic forces of the surging blood, and for connective tissue (commonly encountered as gristle in the hamburger) in general. The matrix to which cells attach in forming organs is rich in collagen. Stable as it is, collagen is destroyed (like practically all other proteins) by boiling: the ordered chains separate, become more or less random coils, and end up as *gelatin* (hence the name of collagen, as in *collage*, from the gelatinous glue prepared by boiling fish skin, and known in gastronomy as aspic).

In connective tissue collagen coexists with *elastin*, a protein of quite different character, though also rich in proline and glycine. Elastin takes the form of a highly flexible random coil. Separate polypeptide chains are chemically linked to one another at intervals through certain of the side chains, to form an elastic net. When made into a fibre, elastin can be stretched at least five times as much as a rubber band without breaking, and it recovers at once when released. In the tissue the elastin net is interwoven with collagen fibres, and thereby acquires unique material properties, matching elasticity with resilience. The walls of arteries consist of concentric layers of muscle proteins, which can contract actively in response to a chemical signal, elastin, and collagen. Collagen also makes up about a third of the weight of bones. The fibres are embedded in a matrix of inorganic material, predominantly a form of calcium phosphate called hydroxyapatite. The bone is thus a composite, hard and shock-resistant, thanks to the crystalline inorganic matrix, yet resilient and capable of modest distortion, thanks to the collagen.

Composites occur throughout nature, wherever toughness and durability are of the essence. The shells of marine molluscs are a good example. That of the Pacific abalone (much sought after by gourmets) has been widely studied. Its principal constituent (as of all others) is calcium carbonate, but it is tougher by a factor of about 3000 than any of the known crystal forms of this compound (which include marble and Iceland spar). In some shells the calcium carbonate takes the form of the mineral calcite, but in abalone the crystals are of aragonite. (All these minerals differ only in the geometrical arrangement of the molecules in the crystal.) Flat blocks of aragonite, their crystal axes all aligned, are separated in the shell by sheets of protein, which guide its initial deposition and function as an adhesive. Because the protein chains are stretched, they afford protection against cracks, a special feature, as we have seen (and will come to again), of composites. An industry has now grown up dedicated to the development of artificial composites from biological or partly biological materials—spider silk threads in silica matrices, and so on—for they unite the virtues of tensile strength, stiffness, and lightness. Many artificial flaw- and crack-resistant, yet plastic, materials have been produced in the laboratory, based on biological models of design, not always with biological materials. Layers of polymer, or even clay, interspersed with hard inorganic crystalline 'platelets'—calcium carbonate, silica, or aluminium oxide for instance—of precisely defined dimensions, are among the designs that can yield remarkable outcomes. It remains to see whether these can be produced in useful bulk.

Of the elastic proteins, the most remarkable is probably *resilin*. In nature it is found only in insects. It is related in its composition to silk, but its properties are far different. The resilin chains are random coils, chemically bonded (*cross-linked*) to each other at intervals, precisely designed to favour large stretch and rapid recovery. To this must be added resistance to wear and tear, for, unlike most other proteins, it is not renewed, and must last the insect all its life. In essence, it stores energy like a wound-up spring, and is therefore found in insects that need to maintain a high power output, as demanded by the wing-beat of flies and bees, and noise generation by cicadas. Most spectacularly, it is a pad of resilin in the rear legs that allow a flea to jump a prodigious distance or to a height equivalent for a man to leaping over Big Ben. With the assistance of resilin, the click beetle can project itself into the air when pursued by a predator: it lies on its back, flexes itself into a jack-knife position, which then springs open to generate the impulse. It has been estimated that the beetle's acceleration is a phenomenal $2300g$. To perform these feats such insects use muscles to compress the resilin pad. Some form of trigger is then activated, the resilin decompresses, and the stored energy is released in one great surge. It has been estimated that the elastic efficiency is about 97%, or in other words only 3% of the energy is lost as heat. This far exceeds the elastic efficiency of the best synthetic rubber. The endurance of resilin is almost as remarkable, for the resilin plug that assists an insect's wing-beat must undergo millions of expansion–contraction cycles during the insect's lifetime. It is therefore not surprising that attempts are in hand to produce resilin in genetically engineered bacteria for such specialized uses as surgical repair of arteries.

Globular proteins: principles of design

A large proportion of the vast repertoire of proteins which our genes make available to us are *globular proteins*, and most of these are enzymes—the catalysts on which our metabolism depends. Globular proteins, which as their name implies are tightly packed, compact objects, cannot accommodate the kind of long regular rod-like polypeptide chains that characterize the fibrous proteins. They do, on the other hand, make use of the folds that the chain, by its nature, can enter—the α-helix, the β-structure, and the turn. Whether a length of polypeptide chain enters one or other or none of these depends on the sequence of amino acids. So where a succession of residues favours the α-helical structure, an α-helix is likely to form.

But a dominant factor in the evolution of globular proteins is the inclination of hydrophobic ('water-hating') side chains—those especially with a hydrocarbon (alkane) chain or a ring—to cluster together, away from water. So the protein globule is

47

often compared to an oil drop—a small sphere that minimizes the proportion of the hydrocarbon molecules in contact with water. This tendency is called the *hydrophobic effect*, and is the main energetic driving force causing a protein to assume its globular form. The trend is assisted by the presence in the protein of side chains that bear a positive or a negative charge, typically around one-fifth of all its amino acid residues. Most of these charges are on amino ($-NH_3^+$) or carboxyl ($-COO^-$) groups at the tips of the side chains in which they occur. Charged species (ions) are soluble in water and have a strong preference for the watery over the oil-like environment in the heart of the protein. Therefore the outer surface of a globular protein bristles with positive and negative charges, which is why proteins will migrate in an electric field towards the cathode or (as most often) the anode, depending on whether negatively or positively charged residues predominate. This, in fact, is the basis for separating the proteins in a mixture by the method known as electrophoresis.

There is, of course, more to it than that, for if one were to generate a random sequence of amino acids, there would be little chance that such separation of hydrophobic and hydrophilic side chains could happen. The hydrophobic side chains in a globular core must pack intimately together as though forming a three-dimensional jigsaw (though with rather soft components). This takes millions of years of evolutionary trial and error to achieve. The three-dimensional structure, or *tertiary structure*, is thus defined (like the secondary structure) solely by the sequence. The structure of a relatively small protein—chymotrypsin, an enzyme that degrades proteins in the pancreas—one of the first to be unravelled by X-ray crystallography—is shown below:

Figure 7 Model of chymotrypsin, a typical small globular protein. This representation is a *space-filling model*, in which each atom is normally represented by a colour-coded ball.

A structural feature worth mentioning pertains only to the important minority of globular proteins confined to fluids outside cells. These include the proteins of the blood serum and of the pancreatic and stomach juices (like chymotrypsin, shown above). To enable them to survive in an unsheltered milieu (hostile even, as in the hydrochloric acid-rich stomach juice), they have acquired an additional source of structural stability: they have one or several *disulphide bonds*. These arise from one of the twenty types of amino acid, cysteine, the side chain of which carries a *sulphydryl* group, —SH. This group is very easily oxidized (oxidation being taken to mean chemical addition of oxygen or elimination of hydrogen; the converse of oxidation is reduction*).

* More generally oxidation can be defined as the loss, and reduction the gain of an electron.

49

Under conditions that favour oxidation a pair of sulphydryl groups, which may be in distant regions of the sequence, can bind to each other by oxidation to form a disulphide bond, $-S-S-$. A bridge is thereby created, which if correctly positioned in the tertiary structure, will greatly enhance its stability.

Another, and quite abundant, category of globular proteins is one not entirely globular. The protein may contain a globular part, with something like a random coil attached to it. Such a molecule can be envisaged as a tadpole. Others have a globular and a stiff filamentous part, or domain. A particularly well-studied example is the muscle protein myosin, about which we will have more to say. It possesses two globular 'heads' attached to a long rigid filamentous tail. Many myosin molecules can pack together to form much longer and thicker filaments, from which the heads protrude. There are also some proteins that function as random coils, or something close to that.

Structural motifs

Globular proteins, then, contain, in varying proportions, stretches of α-helix, β-sheets, turns, and segments of no regularity. The ordered elements in the chain also fold on each other and most often form constellations of structures repeatedly found in quite diverse proteins. Among these (illustrated below, along with some whole proteins) are β-sheets formed when the chain repeatedly folds back on itself, like a concertina. Or often the turns that the chain makes in doubling on itself are replaced by longer segments containing α-helices. At other times, such an assemblage closes on itself to form a barrel shape; this occurs in chymotrypsin, and can be more easily seen in a representation of the backbone alone:

Figure 8 Chymotrypsin structure, showing only the course of the backbone without the side chains.

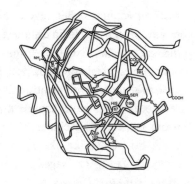

Or again, a series of β-structures may twist themselves into an arrangement that looks like a propeller:

Figure 9 Representation of the small protein, thioredoxin, in a *ribbon diagram*. The twisted parallel β-structure is indicated by the arrows in the centre, with helices on the outside.

The pictures here are of small proteins, as proteins go, and the structures of large proteins, which may be ten times or more the size of chymotrypsin, are much more complex and generally contain a number of different, or similar, structural motifs. There is clear evidence that these and other such 'modules', as we may call them, have been planted in different proteins by evolutionary descent from some primeval protein ancestors. Moreover, as

51

evolution has proceeded, parts of genes have migrated around the chromosomes in which they reside and have fused with other genes. This is why the larger proteins, especially, are most commonly made up of domains capable of forming independent globular structures by themselves; in some cases they may even exercise different physiological functions within the one large protein molecule.

Protein subunits

A globular protein molecule may associate with one or more others of the same or a different kind. The separate elements are then referred to as *subunits*, and the form that the assemblage takes is the *quaternary structure*. This affords a whole new range of functional options, for one subunit may regulate the action of another in response to a structural signal, or the subunits may exert different, but interdependent functions. One of the best understood examples is the red oxygen-carrying protein of the blood, haemoglobin, which contains four oxygen-carrying subunits of two kinds, called α and β, so that the protein can be designated $\alpha_2\beta_2$. When any one of the subunits binds oxygen it undergoes a small but telling change in shape—in tertiary structure. Now the original and the altered forms are structurally incompatible, and cannot fit together, and consequently all the other subunits, whether they have bound oxygen or not, are forced into the new structural state. In this condition the subunits have an increased affinity for oxygen, so in the lungs, where oxygen is plentiful, all four take up their molecules of oxygen and carry them off around the body. When the red blood cells in which the haemoglobin is packed arrive in an organ or tissue in which oxygen is scant and in demand, most of the subunits will have shed their oxygen, and made the transition to the low-affinity structure, and so will give up the remaining oxygen more readily. This kind of interaction between subunits is termed *cooperativity*, and operates in the wide variety of physiological contexts in which such fine control is required.

Many enzymes are made up of regulatory and enzymically active, or catalytic, subunits. The protein, in an inactive state, receives a signal in the form of a generally small molecule, or *ligand*, shown here as a triangle which binds to the regulatory subunit, and in doing so effects a small change in its structure. The altered structure is incompatible with that of the attached catalytic subunit, which consequently undergoes an answering structural change to its active state, as shown here in schematic form.

Figure 10 An enzyme regulation system. The enzymic subunit is pale, and the regulatory subunit dark grey.

The biochemical reaction catalysed by the enzyme is thereby set in train. The molecule on which it acts, its *substrate* (black dot), can enter the open cavity where the reaction occurs. The products are spewed out, and the next substrate molecule can enter. The subunit structure of proteins thus opens a new range of control mechanisms, for cells make use of different signalling molecules (calcium ions, hormones, and many more) to switch on and off the metabolic processes on which life depends. In the complex and highly crowded environment of the cell interactions between proteins are constantly taking place, controlled by signalling molecules and local concentrations of the interacting components. The activities of most proteins are probably controlled in this way, and one of the most active areas in the field of drug research is the pursuit of compounds that will block critical interactions. This requires knowledge of the exact structures of the proteins in question and especially how they interface with each other; the aim then is to design a small molecule which will bind tightly to the area in one of the proteins at which it would attach to its partner.

An extreme kind of quaternary structure prevails when protein subunits bind to each other in something like a head-to-tail manner, for then there is nothing to stop the association from going on and on, resulting in a long string of connected subunits. This allows such proteins to switch from the lone globular to the filamentous state—a common requirement within cells. The transformation is most often provoked by the appearance of a small molecule (ligand) that binds, as above, to the subunit, induces a change to a structure with affinity for another identical subunit, and ends in formation of long strings of subunits. To form stable filaments of globular subunits normally requires multiple intersubunit contacts, and then a helical configuration results. An important protein of this kind is *actin*, a component of muscle (see below) and other complex structures.

A notorious example of continuous self-association occurs in the *prion* proteins, which are responsible for neurodegenerative diseases such as BSE (mad cow disease). An isolated prion protein, a normal constituent of nerve cells, undergoes, for unknown reasons, a switch from its normal to an aberrant structure. This has a high affinity for like molecules, which are accordingly dragged into the aberrant structural state by the rogue subunit, to which they bind. This initiates a wholesale process of filament formation with catastrophic effects on the cell. Or so the process has been envisaged by the researchers who discovered prions; it may in reality, though, be a good deal more complex.

Dynamic structures: the quivering rig

The ability of most proteins to switch between one structural state and another might seems to imply that it is a simple two-state (or multistate) gadget, like say a cuckoo clock, but this is far from the reality. In truth proteins are 'soft' structures, undergoing perpetual, though very limited, fluctuation. Such flexibility plays a part in enzymic activity. An enzyme must respond to its substrate. It must adapt to bind the substrate (usually at a specific site, deep in the protein), to deform it into the state in which it will suffer some form of chemical change, and to release the products of that change. As one of the pioneers of protein chemistry who has measured these fluctuations in protein structure has put it, the entry of a substrate or any ligand molecule into the protein should not be thought to resemble a man entering a room, but rather a cow entering a tent (something he had himself experienced). Fluctuations of protein structure, whether spontaneous and continuous or driven by interaction with other molecules, can occur on a time scale anywhere between milliseconds (thousandths of a second) and picoseconds (trillionths of seconds). Methods exist for observing such processes, and protein dynamics are a particularly challenging area of study.

The folding problem

When a protein solution is heated, or when acid, alkali, or various unfriendly substances are added, the compact protein globule unfolds and turns into a random coil or something close to it. This transition from the *native* (meaning normal or functional) to the unfolded, or *denatured* state is called *denaturation*. In principle a randomly coiled chain, whether in the cell or the test tube, should spontaneously find its way into the energetically most favourable, globular state. This does indeed occur, at least in the case of most relatively small proteins. In such a case denaturation in the test tube is reversible. Heat a solution of the protein, for instance, and it unfolds; cool it and the folded state is regained. But if the concentration of the protein in the solution is high enough that the unfolded molecules can collide and get entangled, like separate lengths of string, sections of the chains are apt to associate tightly in extended pleated sheets (β-structure), as in a boiled egg. The protein chains are thus trapped in a state from which they cannot escape into the folded, native conformation: the boiled egg will not unboil when cooled. Of course, when the polypeptide chain is first synthesized in the cell from its constituent amino acids, it has to find its way into the one unique native state. Just how it does so is a field of study of its own.

For reasons that will become clear when we come to discuss simple polymers like polyethylene, the chemistry of the polypeptide chain allows only two angles of twist between any two successive amino acid residues. All the same, with, say, 100 residues in the chain, the number of possible combinations of all the angles of twist in the chain is 2^{100}; in other words, 2 multiplied by itself 100 times, which works out as 1 followed by thirty 0s, a number of astronomical magnitude. The time needed for a protein chain to explore all these possibilities at random in order to find the one correct combination that defines the native state would be something close to the age of the Earth. Yet some proteins can accomplish the feat in a few microseconds (millionths of a

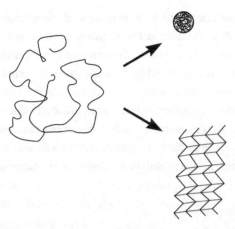

Figure 11 The unstructured, random-coil state of the protein chain folds into the globular state (shaded), or if the concentration is high and the chains can interact, it may from a β-chain (pleated sheet) assembly (lower-right).

second), while others, it is true, can take quite a few minutes. It follows, in any event, that there must be short cuts to the folded states. What in fact appears to happen is that small, relatively stable elements of structure, dictated by the local sequence, form. These guide the folding of neighbouring residues and the parts coalesce into the fully formed globular conformation. Evolution has evidently seen to it that this process is favoured, for sequences that take too long to find the correct tertiary structure can create terrible problems, as we shall see shortly.

Large proteins in general find the greatest difficulty in performing the folding trick, and for the many proteins, large or small, which do not fold efficiently on their own, nature has developed a special class of helpers called chaperones. These are proteins, some very large, some quite small. The members of one class, the chaperonins, are made up of many subunits, which between them create a central cavity. The individual disordered chains go in, and in this sequestered environment they cannot enter into aggregates. But this is not the whole story, for the chaperonins are active folding machines, driven by adenosine triphosphate (ATP). Other, less complex, chaperones, which shield the unfolded protein against aggregation, are the *heat shock proteins*. These are made when the cell is subjected to a rise in temperature or some other form of stress. Such conditions are unfavourable for folding, and therefore call for increased intervention by chaperones.

What do globular proteins do?

We can group the globular proteins into a set of broad functional classes, and evidence is now indeed emerging that proteins of a given class, no matter how seemingly diverse in evolutionary origin or in amino acid sequence, have structural features in common.

Carriers of burdens and guardians of the territory

We might start with haemoglobin and the other proteins that serve only to carry ligands, such as oxygen. There are for instance proteins in the blood which serve to transport vitamins, or metals, or lipids (fats) or hormones, and surrender them where they are needed. Among the proteins designed to take up ligands are the antibodies of the immune system, which all have the same gross structure. Each antibody, though, has its own unique constellation of amino acids—the *binding site*—at its extremities. These will recognize and snap up a particular intruding substance—the *antigen*—specific to that antibody, and eliminate it from the circulation through interactions with specialized cells. The specificity of recognition is established when the antigen first enters the body; an antibody is synthesized, and commonly remains in the circulation for years. This is of course the basis of immunization.

Proteins for seeing

Proteins lie at the root of vision (leaving aside even those that form the lens and other structures of the eye). The chain of the protein, opsin, is built around a cavity which precisely accommodates a small molecule, retinene, closely related to, and derived from, vitamin A. Retinene can exist (like the ethylene derivatives, described on p. 110) in two states, one straight—the *trans* form, one bent into a V-shape—a *cis* form. It is the *cis* form that fits into the opsin to form rhodopsin (from the Greek, *rhodos*, rose), the 'visual purple' of the retina. When a photon—a quantum of light—impinges on the retinene, the chain flips into the *trans* state and falls out of the cavity. An answering change in the shape of the opsin closes off a protein tunnel in the membrane through which calcium ions can otherwise pass. The resulting imbalance in calcium ions sets in train a sequence of events, culminating in the release of a neurotransmitter, the agent that sends an impulse down the optic nerve to the brain.

Rhodopsin is actually the visual protein of the retinal rods, which come into operation in dim light, and do not discriminate colours. Colour vision relies on the cones. These contain a closely related protein called iodopsin (Greek, *iodos*, violet, as in iodine, the element, which forms a deep violet-coloured vapour when heated.) There are three types of cone, characterized by different iodopsins, which respond to light of different colours. Colour-blindness results from hereditary defects in the gene for one or other of the cone proteins. The French word for colour-blindness is *daltonisme*, after the Mancunian chemist John Dalton, whom we have already encountered (p. 2, f.n.). Dalton was famously colour-blind, and outraged his fellow Quakers by sporting his scarlet doctor's gown, which he perceived as grey. He surmised, when he became aware of his condition, that the jelly-like vitreous humour filling his eyeball was blue, rather than colourless as in normal eyes, and so would filter out red light. He therefore gave instructions to his assistant to remove his eyes after death to check. The assistant did as he was instructed, and Dalton's theory was proved wrong. The eyeballs, in a jar, were deposited in the archive of the Manchester Literary and Philosophical Society, and close on two centuries later yielded a sample of DNA, from which Dalton's genetic defect was identified.

Proteins for pumping

Next there are the membrane transporters, which drive ions or sugars, such as glucose, into or out of the cell. These proteins protrude through the fatty membrane that encases each of the body's cells, and therefore have a design of their own: they are composed of hydrophobic segments, compatible with the oily material of the membrane rather than with water, and hydrophilic, sometimes globular, parts, in contact with the watery media on the inside and the outside of the cell. The hydrophobic parts that snake through the membrane (usually several times) are most often α-helices with a length that exactly matches the thickness of the membrane.

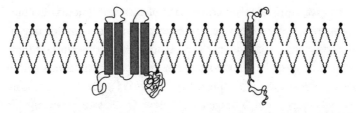

Figure 12 Schematic view of a membrane. Proteins (shaded) are embedded in the membrane, which consists of the layers of lipids. Most of these, the *phospholipids* have a negatively charged phosphate group (black) in contact with water inside and outside the cell, and attached to two long hydrocarbon chains. The membrane proteins may consist of a single α-helix in the lipid, with external and internal domains, or more often, of a series of α-helices, joined by loops, as shown on the left.

Now the composition of the fluid within the cell is different from that outside: the cells, for example, contain potassium ions and practically no sodium, while in the external fluid this balance is reversed. Thus there must be a mechanism for maintaining this imbalance, which is essential for the working of the cell. Similarly calcium ions, which have a wide range of functions in the body, must be admitteded into the cell only when needed and pumped out when they have done their work (for instance, activating a muscle). This is the task of the ion pumps, which push the unwanted ions out and let the desired ions in. The pump protein captures the ion in its binding site, switches the gate to a pathway through the membrane to its open position, and releases the ion on the other side. Ion pumps are in effect engines that work uphill, in that they push material (ions) against the gradient, from a region in which they are sparse to one in which they are abundant, contrary to the natural direction of flow. This needs an energy input, which comes from the consumption of the universal biological fuel, ATP, adenosine triphosphate.* There are also, however, passive channels, which simply allow only molecules of the right kind to flow through. An example is the protein that allows the waste product, carbonate to escape from the cell.

Proteins for signalling

Then there are the signal transducers—proteins such as the membrane *receptors*—which bind some specified molecule on the outside of the cell. This causes a disturbance in the structure of the protein, which in turn activates an interaction on the inside of the cell; the external agent, in other words, causes a chemical signal to pass through the membrane and into the cell. Hormones act through such receptors. Thus insulin, released from the pancreas, binds to its receptors on various kinds of cells, sending a signal to the machinery within to stop release of glucose into the blood. In the cell a cascade of signalling events, each step regulated by a dedicated protein, is initiated, the end result of which is to switch on or off the activities of the enzymes that provoke or shut down a metabolic process.

Proteins for reproduction

This class comprises the proteins that bind to nucleic acids, to DNA or its close relative, RNA. They regulate at different levels the activity of genes. So a protein may, by binding to a designated region of the DNA, untwist the structure, and thus render a particular gene or a set of genes acting in concert accessible to

* ATP drives all energy-requiring processes in all life forms. We consume and synthesize our body-weight of ATP everyday.

a second protein. A chain of events thereby initiated terminates in the formation of the protein encoded by that gene. Or, contrariwise, proteins can cause that gene to be silenced. Proteins copy the DNA when cells divide and multiply, and when progenitor cells give rise to identical (or very nearly identical) progeny during reproduction. A battery of proteins is on hand to repair damage to DNA caused by chemicals or radiation, and proteins proofread the newly formed DNA copy and put right any mistakes. Proteins of course also operate the complex machinery of protein synthesis, starting from the free amino acids. There are proteins that package yards of DNA or RNA of viruses, or the DNA of sperm, into parcels perhaps a few micrometres in size, and some that twist DNA helices into pretzel shapes, while others allow them to untwist again.

Proteins as motors

These are the proteins that cause the contraction of muscles and make cells divide. To carry out such functions they need an energy source, namely (again) ATP. The myosin of muscle has the form of a long rigid rod comprising two α-helices twisted into a coiled-coil, both terminating at one end in a compact globular 'head'. In the muscle the myosin rods are assembled into thick filaments from which the heads project at regular intervals. Interspersed between these filaments in regular array are thin filaments composed of the ubiquitous protein, actin. The actin filaments are two-stranded helical assemblies of more or less spherical subunits (p. 52). The myosin heads are enzymes: they contain sites at which the ATP, which fuels muscle contraction, is 'burned': one of its three phosphate groups is split off, leaving adenosine diphosphate, or ADP. With ADP attached in place of ATP, the head acquires an affinity for actin while also undergoing a shape change, so that it swivels, pulling the actin with it. With ATP in plentiful supply, the bound ADP is displaced by

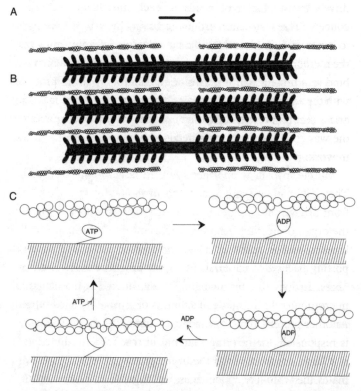

Figure 13 Diagram of the arrangement of proteins in a muscle fibre. A. myosin, with its rod and two globular heads. B. lattice of myosin filaments, formed by packing the myosin rods together, with the heads protruding as shown. C. between the myosin filaments, the actin filaments made of globular subunits (open circles).

ATP once more, the head falls away from the actin and reverts to its previous orientation, and the four-step cycle (as indicated in the diagram) begins again, moving the myosin along another notch (Fig. 13C).

This of course all happens very quickly, and contraction is driven by thousands of heads in each muscle fibre acting in concert. The instruction from the brain, by way of the nerve, to contract is implemented by the release of calcium ions from the membrane that surrounds the muscle fibres. The calcium ions bind to a protein that is part of the regulatory system associated with the actin filaments. Once again a structural distortion ensues, and a second protein component of that system is moved out of the way to allow the myosin heads to bind to the actin and start to work.

Muscle myosin is one member of a large family of myosins. They differ in size and shape and in the actions they perform, from driving the migration of cells to operating moving parts in the organ of hearing. Another important motor protein is called kinesin, and is active in a variety of processes, such as transporting packets of materials ('cargos') up immensely long nerve fibres. It does this by travelling along, not actin filaments, but microtubules—rails made of subunits of a protein called tubulin. Another protein, dynein, also travels along the microtubules and is responsible for carrying material in the opposite direction. It is a large, complex protein with many subunits, which also animates the whip-like appendages, the flagella, that allow bacteria to swim. Dyneins also occur in more elaborate structures such as the cilia, which exert the same function in many micro-organisms, such as the tiny creatures found in ponds and rivers, as well as in mammalian sperm.

Antifreeze proteins

A remarkable class of proteins are the antifreezes. Arctic fish subsist at temperatures well below the freezing point of water. If their blood were to freeze the consequences would be dire. Why it does not is something that baffled marine biologists for at least a century, until it was discovered that the blood contains a special protein. Since then antifreeze proteins have also been found in insects and crustaceans, and in some plants. They differ widely in structure. Some are no more than a single α-helix, while others contain predominantly β-structures. But the feature they all share is that they are made up of repeating groups of amino acids, apparently spaced in such a way as to form hydrogen bonds (p. 24) with the water molecules arrayed along the face of an ice crystal. In this way they prevent the growth of the crystals, and no mass of ice is formed. Many practical applications for these proteins (for the benefit of man, rather than of fish or insects) have been envisaged. Among them are the creation of genetically modified crop plants equipped with antifreeze proteins, so as to improve their frost-resistance; protection of foods and other perishable commodities against damage through freezing; and limitation of tissue damage in cryosurgery. The genes for antifreeze proteins can be introduced into bacteria for manufacture in bulk, and it has even been proposed that they might be incorporated into a spray to prevent ice formation on aircraft wings.

Housekeeping proteins

Finally we must consider the myriad of enzymes needed to run the metabolism and maintain life, and which occur in all, or nearly all, cells. Enzymes are catalysts, that is to say, agents which accelerate chemical reactions. Without a dedicated enzyme most reactions in the body would proceed so slowly as to amount to stasis. Enzymes accelerate reactions by factors of tens to hundreds of thousands. The action occurs at a site on the protein surface, or in a cavity in the structure, referred to as the *active centre*, made up of a constellation of amino acid side chains, tailored to the shape of the molecule on which the enzyme acts, its substrate. The substrate binds at the active centre, and is held in place, in some cases in a state of tension that will cause it to split; or it may be presented in its immobilized state to another molecule with which it is to react. Of enzymes there is practically no end. There are those which break things down—proteins, nucleic acids, sugars, polysaccharides—and others which synthesize. There are enzymes that will attach something to another enzyme by a covalent bond to change its activity, most commonly a phosphate group, $-PO_3^{2-}$. Others again will bring about oxidations or reductions (p. 49). Some will change single into double bonds, some the opposite. Some will lift groups of atoms from one part of a protein and insert them elsewhere in the same molecule or in another one of the same or a different kind. There are enzymes that work under conditions in which other proteins would be denatured. The most spectacular examples occur in the thermophiles—bacteria which thrive in hot springs or in thermal vents under the ocean at temperatures close to boiling. Such heat-resistant enzymes are used in biological washing powders.

Mighty molecules: enzymes revealed

The study of enzymes and how they work began seriously in the nineteenth century, and well into the twentieth there were influential chemists who refused to accept that these catalysts were indeed proteins. In the late years of the eighteenth century it had already been shown that stomach juices could dissolve meat, and many more such experiments followed over the next hundred years. An American army doctor even studied the digestion process in a living laboratory, the stomach of a victim of a gunshot wound whom he had tended. The wound in the midriff had healed, leaving an aperture into the stomach, through which pieces of food could be inserted for observation. Then in 1897 a German physiological chemist, Eduard Buchner, prepared an extract from yeast cells, which proved to be capable of engendering fermentation. For this Buchner received the Nobel Prize, and *Ferment* entered the German language as the term for an enzyme. Previously it had been generally supposed that processes such as fermentation could occur only within living cells, where there resided a mysterious essence—the *élan vital*. Determined attempts began to track down the source of the catalytic activity in cell extracts, which seemed to be always associated with proteins. The most tenacious of the pioneers was the one-armed American biochemist James Sumner. (He had lost an arm in a hunting accident when he was young.) He set about purifying an enzyme, called urease, from the jack bean, and the more he purified the protein, the more activity he got. Another American biochemist, John Northrop, was making the same observations on stomach enzymes that digest proteins.

Their conclusion, that enzymes are proteins, was derided by leading European organic chemists, chief of whom was the illustrious German Nobel Laureate, Richard Willstätter. The view, prevalent especially among the German chemistry establishment, that proteins were ill-defined colloidal 'smears', died hard, and Willstätter was apparently predisposed to believe that such powerful chemical activity must reside in smaller, better-defined compounds. And indeed, the more he and his students purified, the less protein they appeared to have. It was the great activity of their enzyme (from saliva) that deceived them, for they were working with minute amounts of material, and they found activity where they could detect no protein, because there was so little of it. The furthest Willstätter was prepared to go was that proteins might in general be the carriers of catalytic compounds, more or less loosely attached to them, and in 1920 he thought he had isolated the active entity. He obdurately rejected Sumner's evidence, and invited himself to Sumner's laboratory at Cornell University, hoping to find fault with his work. But then in 1926, after a decade of patient toil, Sumner announced the crystallization of urease. It was taken as read by chemists that to crystallize a substance must be pure. Four years later, Northrop crystallized the digestive enzyme pepsin.

Some years earlier, in 1924, Willstätter's illustrious career had come to an abrupt and unhappy end. Willstätter was a Jew in an anti-Semitic environment. The Faculty of Munich University met to appoint a new chemistry professor, and Willstätter put forward the name of Victor Goldschmidt, the world's leading geochemist, who had also been nominated by the retiring occupant of the professorial chair. Immediately an angry mutter arose round the table: 'Another Jew!' Willstätter rose, left the room, and wrote a letter of resignation. He was, at fifty-three, at the height of his powers. Despite a petition from students and many of the Faculty, he would not be coaxed back, rejected all other offers, and withdrew to his house. His successor, however, offered laboratory space to Willstätter's assistant and former student, who continued to work under the supervision of her patron, with whom she conferred on the telephone, but never again met. This curious arrangement persisted until Willstätter was finally hounded out of the country by the Nazis. He died in penurious exile in Switzerland in 1942. In 1946, Sumner, Northrop, and Wendell Stanley, who had crystallized the tobacco mosaic virus in 1935, shared the Nobel Prize. Enzymology, the study of the chemistry of enzymic reactions, remains a thriving science, some of it now directed towards the *ab initio* design of proteins that will catalyse the particular reactions that chemists desire.

When proteins go wrong

It does not take much to cause physiological mayhem. Mutations or chemical damage can bring this about. A mutation in general results, as we have seen, from a random change in one base (see below) in a gene. This may happen through an error of copying the DNA when a parent cell gives rise to its progeny, or it may be caused by an extraneous event, such as exposure to X-rays or ultraviolet light from the sun, or to a deleterious chemical. Or of course an already established mutation is apt to be transmitted from parent to offspring. Most mutations are harmless or relatively neutral, and some—a very few—will turn out to confer an advantage. Those are the ones that possess survival value, and will tend to gain ascendancy in the population. Other mutations are disadvantageous or even lethal, those for instance that prevent the production of an essential protein, or give rise to a protein which does not function properly. One of the best known examples of a deleterious mutation transforms normal into sickle-cell haemoglobin. A change in one base in the DNA, with the corresponding change of one amino acid in the protein stemming from that gene, results in a haemoglobin that is perfectly normal in respect of oxygen uptake and release, but possesses a new and catastrophic attribute. When it unloads its oxygen the changed amino acid, exposed on the exterior of the two identical β-subunits of the $\alpha_2\beta_2$ haemoglobin (p. 52), causes the haemoglobin molecules to attach to one another. They do this so effectively that long, thick filaments of haemoglobin form in the cell and irreparably damage the cell membrane. The cell distorts into a characteristic curved (sickle) shape, and loses its elasticity, so that it can no longer squeeze through the capillaries. These become blocked with dire effects.

Why then, you may ask, has this mutation, which in earlier days cut short the life of a victim before puberty, survived in the population? For one thing, one copy of the haemoglobin gene (one allele) is contributed by each parent. Suppose now that both have one normal and one mutant allele—they are heterozygotes, or carriers, and only mildly affected by the mutation. Then, by the laws of heredity, each child will have a one-in-four chance of ending up with two mutant alleles—(s)he will be a homozygote, possessing only the mutant haemoglobin—but also the same chance of inheriting only the normal alleles, and so become a normal homozygote, making only normal haemoglobin. But the child will have a two-in-four chance of ending up a heterozygote like both parents, with the same mixture of mutant and normal haemoglobins. This latter condition is called sickle-cell *trait*, and does not cause severe disease. Geneticists write the heterozygotic condition as A/S (A for normal haemoglobin of the human adult, and S for sickle). Then recombining the alleles of two heterozygotes, A/S and A/S, we may get A/A, A/S, S/A, and S/S. A/S and S/A are the same, regardless of which parent contributes the S and which the A; hence the statistical ratio of probabilities of 1:2:1 of these states. Now, there is good evidence that in Africa, where the sickle-cell mutation arose and survived, and where there is even now a high incidence of malaria, the

71

sickle-cell haemoglobin gives a degree of protection against the parasite. And so, presumably, over the centuries more lives have been saved from this benign effect of sickle-cell trait than were lost by the death of those with full-blown sickle-cell anaemia. (Modern treatments allow homozygotes to survive into at least middle age.)

There are innumerable other examples of mutations leading to disease, most often in enzymes, which may cease to function or may even become hyperactive. A large number of the drugs now marketed, or under development by the pharmaceutical industry, are inhibitors of particular enzymes. They are most often small molecules, closely resembling the enzyme's natural substrate, which will fit into the active cantre and impede access to the substrate.

In recent years there has been much agitation about diseases of 'misfolding'. If a mutation occurs which prevents the protein chain from folding up into its normal globular state the first consequence is likely to be the lack of the activity associated with the protein. A chance misreading of the gene sequence by the machinery of protein synthesis will sometimes occur in any case, but in such cases, an elaborate mechanism comes into play for annihilating these useless, and possibly toxic copies. Enzymes first mark the aberrant protein for destruction by attaching a marker, another polypeptide, called ubiquitin. This label is recognized and taken up by a huge multiprotein complex—a proteasome— containing proteolytic enzymes (proteases); these break the captive protein into small fragments, which are soon eliminated from the cell. This search-and-destroy system may have to cope with quite a large part of the proteins synthesized in the cell. Defects of folding can arise in many ways—from a random error during the complex manufacturing process, from a mutation in the gene, or from environmental stress in the form of a rise in temperature or acidity, or from the incursion of a hostile chemical. A folding defect, if it does not prevent folding altogether, will lower the stability of the globular fold, so a balance will prevail between molecules in the folded and the unfolded state. Unfolded polypeptide chains, if they evade the attentions of the proteasomes, become a danger to the cell, and indeed to life, for they possess, as we have seen, the propensity to form large aggregates, mainly through β-sheets involving sequence elements in separate molecules, rather as shown on p. 57. The appearance of the aggregates varies somewhat, depending on the proteins from which they are formed. Here is an electron microscope picture of typical *amyloid* fibrils:

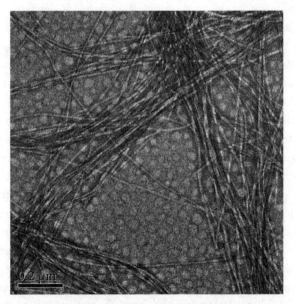

Figure 14 Electron microscope picture of amyloid fibrils, of the kind seen in the brains of victims of Alzheimer's disease.

Once a few such unfolded molecules have clustered together in this way, the aggregate will trap any additional unfolded molecules in the vicinity. They act, in fact, as 'seeds' to initiate something like a crystallization process (as when a crystal of a compound is dropped into a saturated solution of the same compound). This is the way it works: the unstable protein fluctuates between its folded and unfolded states; even if it is in the unfolded state only, say, a thousandth of the time (in other words, at any instant one in a thousand of the molecules is unfolded), this may still be long enough for it to fasten itself onto a similar unfolded molecule, and by this means the aggregation process continues. Such long protein filaments signal trouble for the cell. These are called *amyloid fibrils*, and are an accompaniment, or more probably the cause, of a wide range of diseases. Fibrils of this kind occur in the brains of victims of Alzheimer's disease, Parkinson's disease, Huntington's disease, and many other conditions least to be desired. These conditions are most often an accompaniment of old age, when, it is thought, the body's quality control mechanisms may start to fail. For instance, mutations may accumulate because of impaired proof-reading of the sequence of newly formed DNA, or a reduced supply of chaperones may allow misfolded proteins to escape detection and elimination.

The *prion* diseases are a special case, caused, as we have already seen, by an aberrant form of the normal protein (p. 54). When ingested the abnormal prions subvert the normal population, and by a mechanism still debated, and possibly involving other players, induce them to flip into the abnormal state and generate amyloid bodies. The first prion disease to be traced to its source was *kuru*. The word means 'trembling with fear' in the language of the Fore people of Papua New Guinea, which describes its symptoms. It affects the brain and nerves and is always fatal. An American doctor, Carleton Gajdusek, was awarded the Nobel Prize for Medicine in 1976 for the discovery that the disease was transmitted by cannibalism. The custom of the Fore, at least until the danger of the practice was impressed on them, was to eat parts of their dead ancestors. The men would feast on the flesh, the women and children on the brain and it was they who mainly got the disease. What made it so difficult to identify this as the cause of the diseases was the very long incubation time, for the symptoms did not appear for years, sometimes even many decades later.

Following Gajdusek's detective work in the highlands of Papua New Guinea, the penny dropped and several baffling human and animal afflictions were recognized as infectious prion diseases resembling kuru in their effects. They include scrapie in sheep and BSE (mad cow disease), Creutzfeldt–Jakob disease of humans, and many other neurological conditions, all of them fatal, and characterized by amyloid fibrils and long incubation times. The study of these diseases has contributed to the recognition that the folded state of most proteins is of its nature more or less unstable (to an extent that varies widely between different proteins) And so the threat of amyloid formation always hovers; what evolution has given us is protein molecules of modest stability (thus pulsating and mobile (p. 55), to the extent that their functions demand), yet not so unstable as to give rise too easily to amyloid fibrils. In fact evolution has probably seen to it that the sequences most apt to form β-structures are excluded, so far as the demands of function allow, from the composition of our proteins. This may be seen, of course, as an example of flawed biological design, to add to the many (like the unsatisfactory position of our noses, which would be more efficient if sited in our foreheads, as enumerated by the great J. B. S. Haldane, and other biologists).

Designer proteins

There is now a vast database of protein sequences, and the function of a newly discovered protein can often be deduced from its sequence, if a computer search reveals a similarity to proteins of known function. One can also infer from the sequence, with tolerable but not complete accuracy, the secondary structure (p. 37), if not by analogy with related proteins then by a set of empirical rules relating the two. But the science has not yet advanced so far that one can divine the tertiary structure—the globular fold—from the sequence alone, and conversely, one cannot yet design a sequence to generate with any fidelity a desired molecular architecture. It is possible, though, to modify known sequences in such a way as to alter properties. One can, for example, make a protein by synthesis in bacteria or in animal cells that combines two functions. The sequence of an enzyme can be fused to that of a fluorescent protein (commonly the so-called green fluorescent protein, found in a jellyfish), so that when it is introduced into a cell by injection, or into an animal by genetic engineering, its location and progress can be traced with a fluorescence microscope. This subterfuge has been used to make luminous fish or mice for research. Another such device is to fuse the molecule of interest to luciferase, the protein responsible in nature for the flashing of fireflies. Or one can fuse an antibody to a toxin (a protein lethal to cells), so that this agent can be aimed at a chosen target—a cancer cell, for example, carrying the antigen recognized by the antibody.

A less ambitious aim is to modify an enzyme so as to increase its resistance to heat. The structures of many of the remarkable heat-resistant proteins from the bacteria which live in hot springs or thermal vents under the oceans (p. 67) have been determined, and the features responsible for their extraordinary properties have emerged. Such information allows a fair guess at ways to increase the heat stability of an enzyme of choice. Better biological washing powders have been one result as mentioned earlier. Rudimentary proteins have also been constructed *ab initio*, more in pursuit of the rules of folding than (at least so far) of practical applications. Based on the information gleaned from the structures of natural proteins, such 'designer' molecules have been synthesized to span lipid membranes (p. 61), others to associate into helical coiled-coils (helices twisted into ropes with two, three, or four strands), and so on. Research on designer proteins proceeds apace, and will undoubtedly lead to useful products in the next few years. The general solution to one of the major problems remaining—how to deduce the structure of a protein from the sequence—probably depends on greater computer power, which seems now to be on the horizon.

4

STORAGE: OF FOOD AND INFORMATION

Carbohydrates—seeds, livers, and joints

Carbohydrates are polymers made up of chains of sugars. Sugars are ring compounds and in nature there are two simple kinds: the *pentoses* (the best known of which is fructose, or fruit sugar) have five-membered rings, while the *hexoses* (such as glucose*) have six-membered rings. These serve as building blocks for more complex sugars. Thus sucrose (cane or beet sugar) is made up of one glucose and one fructose unit linked together. These are illustrated

---

* Historically (probably originating from the Greek work, *glykus*, meaning sweet) the ending –ose designates a sugar. When, in 1928, the waggish Hungarian biochemist Albert Szent-Györgyi isolated the anti-scurvy factor (vitamin C) from green peppers, he mistakenly took it for a sugar and wanted to name it ignose; when the editor of the *Biochemical Journal* demurred, Szent-Györgyi came up with godnose. The editor stood firm, and it became hexuronic acid, later ascorbic acid, as it remains.

below in the conventional manner, with the carbon atoms at the corners of the pentagons and hexagons not shown:

D-glucose

D-fructose

Sucrose

The sugar units are referred to as *saccharides*, and nature has many examples of di- tri-, and tetrasaccharides, as well as large polymers, or *polysaccharides*. Natural sugars have asymmetric carbon atoms (p. 20), and so are optically active. Like other optically active compounds (p. 20), they may be laevo-rotatory (L), rotating the plane of polarized light to the left, or dextro-rotatory (D), and rotate it to the right. Natural glucose and natural fructose are D-glucose (but switching —H and —OH somewhere in the ring turns it into a different form, which, depending on where the switch is made, may be D or L).

The first carbohydrate to be recognized (but not, as we have seen, by everyone) as a giant molecule was starch, or rather a water-soluble component of starch. Starch is the insoluble carbohydrate store of most plants, which, when we eat and digest it, is broken down into its glucose units. In hot water starch separates into a soluble polymer, amylose, and an insoluble residue, amylopectin, in which the glucose chains form branches radiating in all directions. Amylose was the second macromolecule (after haemoglobin) to be allocated a molecular weight. In 1888 two English chemists, using a recently developed method of estimating the sizes of molecules, came up with a value of 32,000. This is far below the true size, but it was taken to indicate that amylose indeed had polymeric character. This of course did not at the time satisfy the colloid school.

There are many other polysaccharides, of which the most widely distributed is the material of plant cell walls—of wood and of cotton. This too is made up of glucose units: *

cellulose

Related polysaccharides are the hemicelluloses, which most often occur in association with cellulose. They include xylan, which is abundant in plant cell walls and algae, and chitin, the main substance of insect cuticle. Glycogen, sometimes referred to as animal starch, resembles amylopectin, being a huge, highly branched polymer of glucose units, looking something like this:

---

* We cannot, of course, make use of cellulose for nutrition, but this is exactly what grass-eating animals do. In the USA cows are commonly fed on maize; paper and cardboard pulp, which is nothing more than cellulose, is added to the feed. This is evidently caviar to the cow, and perfectly well digested.

Figure 15 Glycogen is a highly branched chain of sugar units, represented by the black dots. This is the form in which we store glucose.

Humans and other animals store carbohydrate in the form of glycogen in the muscles and liver. It is from glycogen that glucose is mobilized as needed, by enzymic breakdown. Glucose is a central player in our metabolism: our organs, and in particular the brain, require a continuous supply, which comes mainly from glycogen in the liver. Insulin, a small protein, is the hormone responsible for regulating the glucose supply. If the regulatory system fails, diabetes and general metabolic derangement result. Sucrose in food is broken down by enzymes into glucose and fructose. Excessive amounts of fructose cause especial problems, which is why high-fructose corn syrup (HFCS) has stirred up so much concern among nutritionists. This sweetener turns up in almost all soft fizzy drinks. It is produced from maize (genetically engineered in recent years to maximize sugar yield) by a process devised by Japanese chemists when maize was a drug on the market. The starch is extracted and broken down by digestive enzymes, and another enzyme is used to convert the resulting glucose to fructose. The advent of HFCS allowed a well-advertised increase in the volume of Coca-Cola sold per penny, with no diminution in profit margin, and a modest contribution to the ill-health of the population.

Another class of polysaccharides embraces the long, unbranched mucopolysaccharides ('sticky polysaccharides'), a term that has now been formally replaced by glycosaminoglycans (GAGs). One such is hyaluronic acid, composed of thousands of sugar units, an important lubricant in the joints. It is also found in connective tissue and in the matrix to which the cells of the body's organs are attached. Other GAGs, such as chondroitin sulphate, in which a proportion of the sugar units carry negatively charged sulphate groups ($-SO_3^-$), are abundant in cartilage and in the cornea, and are also found in arteries, in skin and in bone. GAGs act as lubricants: since they are highly hydrophilic and swollen with bound water, as well as carrying high negative charge, the molecules repel and slide past each other. So the thick solution forms a kind of expanded 'foam', which cushions and lubricates joints and other parts. GAGs also appear to play some part in the movements of cells, and in other dynamic processes, such as wound healing. Joint lubrication is vastly better than anything man-made.

Marine algae, particularly seaweed, are a rich source of gelatinous polysaccharides. Agar, the best known product, is a mixture of two polysaccharides, agarose and agaropectin, both composed of other sugars than glucose. Agaropectin, like chondroitin sulphate, carries negatively charged sulphate groups. Some species of algae also yield up another gelatinous sulphated polysaccharide, called carrageenan, containing different sugars again. All these polymers are components of the cell walls, and have a vast range of applications in the laboratory and more especially in industry. They are components of all manner of unctuous foods—ice-creams, mousses, bottled sauces, and the like. All of these are emulsions (suspensions of fat droplets in water), and the polysaccharide gelling agent is added as stabilizer to prevent the emulsions from separating (like home-made mayonnaise when the oil is added too quickly). Semi-synthetic polymers, such as methylcellulose, are equally widely used: they are most often the soft unattractive exoskeleton that remains, quivering in the dish, when a mass-produced ice-cream is allowed to melt. These polymers also serve to thicken creams, sauces, some paints, and many cosmetics. The mutually repellent 'bottle-brush' polymers, (see below), are also stimulating attempts to create water-based lubricants to replace oils, where those are undesirable. This scheme works best on soft surfaces, which can deform under mechanical pressure, like cartilage, but that of course limits the scope of industrial applications.

Glycoproteins and proteoglycans

Glycoproteins are found in all life forms. They are formed by enzymic attachment of sugars (a process called glycosylation) to a newly synthesized protein on specified amino acid side chains. The sugars are most often in the form of a short chain (seldom more than a dozen or so sugar units), emanating like whiskers from the globular protein. The sugars are involved in recognition processes between cells. Among the most familiar glycoproteins are the blood group substances, which cover the surface of red blood cells. There are four major blood groups and some twenty known minor ones. Our blood group types are inherited from both parents. Transfusion with a blood of the wrong group can be fatal when antibodies in the recipient's blood plasma identify the new cells as alien and react with them. (Antibodies themselves also have a few attached sugars.) The reason for rejection of incompatible organ or skin grafts is that the body's immune system recognizes as alien the *histocompatibility molecules*—glycoproteins of a special class—on the graft surface, and then attack and destroy the intruder.

Proteoglycans differ from glycoproteins in that they consist of a single large structureless protein chain to which are attached at frequent intervals long chains of GAGs (see above), often comprising several hundred sugar units. The resulting molecule can have a molecular weight in the millions, the bulk of which (up to perhaps 95%) is carbohydrate. The structure has the form of a rather floppy bottle-brush, in which the wire backbone is protein, and the bristles are chains of sugars:

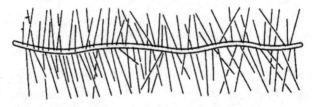

Figure 16 A Proteoglycan molecule, shown here, consists of a long protein chain (the core) with attached polysaecharide chains, protruding like bristles, and making up the bulk of the molecule.

Proteoglycans are found in the net (extracellular matrix) to which cells adhere, and especially in connective tissue, where they form a soft, highly charged, water-retentive, gel-like matrix within which collagen fibrils are embedded.

The nucleic acids—genes to computers

DNA

The history of DNA, deoxyribonucleic acid, the world's most celebrated molecule, began inauspiciously: it was first prepared in 1869 from pus by a Swiss doctor, turned physiological chemist, Friedrich Miescher. Looking for new proteins, Miescher found that the white blood cells—pus in short—contained a material that seemed not to be protein at all. On looking into its origins further, Miescher observed that it was confined to the nucleus of the intact cell, so he gave it the name nuclein. Its chemical nature emerged little by little over the years, but its true interest became apparent only in 1948, when the work of Oswald Avery and his colleagues at the Rockefeller Institute in New York proved, at least to the satisfaction of the most discerning observers, that deoxyribonucleic acid (as it was by then called) is the repository of genetic information. They had shown that the 'principle' which, when transferred from a virulent to a non-virulent strain of a pathogenic bacterium, imported the virulence was indeed DNA.

It should be said that most biochemists greeted this startling information with derision, disbelief, or at best apathy. It was already known that DNA was made up of only four kinds of bases, so how could anything so rudimentary encapsulate the huge amount of data implicit in the design of a living creature? Surely, they thought, it could only be proteins, with their

85

twenty amino acids, that could give rise to the huge variety of sequence combinations of all the other proteins of the body. It was colleagues of Avery's at the Rockefeller Institute who most vociferously impugned his work, insisting that his pure DNA preparations must be contaminated with protein.

Nevertheless, in some enlightened circles work on DNA continued. As early as the 1940s the organic chemist Alexander Todd in Manchester (afterwards in Cambridge) had determined the chemical structure of the nucleoside constituents (see below) of DNA and RNA. (Todd received the Nobel Prize in 1957, followed soon after by his apotheosis as Lord Todd of Trumpington, or to his contemporaries, Lord Todd Almighty.) Without these structures there could have been no rational basis for building molecular models. Then in 1953 came the *coup de foudre* when James Watson and Francis Crick in Cambridge deduced its three-dimensional structure, the famous double helix from the chemistry of the nucleotides and X-ray diffraction data obtained by Maurice Wilkins and Rosalind Franklin at King's College in London. (Todd sought to put them all in their place: *he*, Todd, had determined the structure of DNA, they, only its conformation.)

Watson told the story of the discovery in his outrageously candid and captivating book, *The Double Helix*. With the self-confidence and ambition of his twenty-three years, he had formulated the modest plan to uncover the basis of heredity. He was in no doubt by then that DNA was indeed the genetic material, and at a meeting in Naples he had been transfixed by an X-ray diffraction picture of DNA fibres, which Maurice Wilkins had shown. It revealed the semi-crystalline nature of the DNA, and convinced him that the structure of DNA—of the gene, no less—could be cracked. Like many others of his generation, he had fallen under the spell of the famous Austrian theoretical physicist Erwin Schrödinger, who had published his cogitations on the nature of life in a slim volume with the provocative title *What is Life?* In it he had asserted, on the basis of specious reasoning it must be said, that the gene must be crystalline. The book had persuaded many physicists and chemists, among them Watson's patron, Max Delbrück, that the central biological problems could be tackled by physicists without loss of dignity.

The DNA chain, it was known, consists of aromatic bases,* of which there are four kinds, A, G (both purines, as on p. 17), C and T (both pyrimidines), attached to a backbone, comprising units of a sugar, a pentose called deoxyribose, and phosphate groups. The combination of base and sugar is called a *nucleoside*, while the base with a sugar phosphate is a *nucleotide*. If we represent the base (A, G, C, or T) by B, we can write the formulae as:

nucleoside   nucleotide

---

* Recalling that a base is the counterpart of an acid in that it can take up (rather than lose) a hydrogen ion, H^+, or give rise to a hydroxyl ion, OH^-, it is plain that we are dealing here, as is most common in biochemistry, with weak bases. This means that they are uncharged under neutral conditions, and acquire a negative charge only in solutions of high alkalinity. So the fact that they are bases hardly matters. But note that phosphoric acid is a rather strong acid, and so the phosphate groups in the DNA backbone are all negatively charged—have given up hydrogen ions, in other words—in the neutral milieu of the cell.

Here then is the structure of the DNA chain:

To form the famous double helix the bases along the sequence (marked B_1, B_2, B_3) must pair off by hydrogen bonding, and the only pattern of hydrogen bonding compatible with the constraints of the helical structure is of A with T, and G with C. One of they keys to the discovery of the double-helical structure was that the A—T and the G—C base pairs have the same shape:

A :::::: T

G ⠿ C

The function of the DNA, as all the world now knows, is to store information. Because of the *complementarity* of the bases, the two strands in the double helix (also called a duplex) are related rather in the manner the positive and negative of a photograph: an A in one strand is opposed to a T in the opposite strand, and vice versa, and a C by a G. The result is of course the now emblematic structure

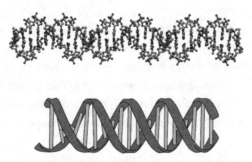

Figure 17 The DNA double helix. The steps on the ladder are base pairs (left) and atomic model (right).

One strand can then act as a template for the synthesis of a new strand from a new set of nucleotides. This new strand will then be identical in sequence to that of the original partner strand of the template. In this manner the offspring receives the genes of the parent. A reason for scepticism about this model of heredity in some quarters was the perceived difficulty of untwisting the long double helix. In actuality this happens progressively, as enzymes work their way along the DNA thread. The biologist J. B. S Haldane saw the point at once when he came to view the model in 1953: 'What you need', he observed, 'is an untwiddlease.'

The DNA in our cells is indeed long: the total extended length in any one cell is about 2 m (but this is divided between 23 separate chromosomes, arranged in pairs, thus making up 46 chromatids, and varying in length from about 3 to about 7 μm. The DNA is thus tightly packaged so as to occupy something like a ten-thousandth of its extended length. The genes are distributed along the DNA chain, interspersed with enormous lengths of sequence with no known function (commonly called 'junk DNA', and consisting partly of defunct genes, the detritus of evolution—but there has been speculation about possible uses for the 'junk').

The DNA molecule is by no means a rigid structure. Even before the double helix was discovered it was known that solid fibres of DNA changed from what was called the A-form under conditions of low humidity to the B-form when wet. The fundamental structure—the right-handed double helix—remains, but its twist alters, and the base pairs tilt relative to the long axis. Something different occurs in those (rare) parts of the DNA in which the sequence...GCGCGC...appears, for then the double helix switches to a left-hand screw sense. This is called Z-DNA, and such regions must be extremely sparse. They may serve as recognition sites for one or other of the many proteins that interact with DNA in the cell. Some sequences of bases also have the tendency to bend the double helix, and the same effect can arise when some proteins attach themselves. This indeed is how DNA, distorted by imposition of a twist on the double helix, allows itself to be packaged into a small volume. Moreover, we have in our cells autonomous bodies called mitochondria. These are thought to derive from bacteria, which at some early stage in evolution took up residence in our cells. They now fulfil an essential role as the cell's energy generators, producing the molecules which fuel our metabolism. The mitochondria have their own DNA and their own genes (which are inherited entirely from one's mother). The mitochondrial DNA is small and has no ends: it is circular. Such circular DNA forms are also seen in viruses, bacteria, and certain micro-organisms. Circular DNA is often twisted into supercoils, for if the primary helix is tightened a little, or loosened by twisting in the opposite direction, the structure will compensate by forming higher order twists. Imagine twisting a loop of string beyond the point at which it becomes tense: the helix will twist into a higher order coil as does a helical telephone cable when it is replaced on the receiver each time by the same left- or right-handed rotational movement of your hand.

Figure 18 Supercoiled DNA.

In the case of the twisted circular double-helical DNA, the introduction of a nick anywhere in one strand (readily done with

91

an enzyme) will cause the tension to relax, the super-twists to disappear, and the structure to revert to a circle.

RNA

The other type of natural nucleic acid is ribonucleic acid (RNA). In chemical terms it differs only trivially from DNA, in that the backbone sugar (ribose) contains one more oxygen atom than the deoxyribose of DNA, and the base, T, is replaced by U, or uracil, which is merely T lacking one methyl ($-CH_3$) group in the pyrimidine ring. RNA can form a double helix, just like DNA, and does so in a class of viruses, and in some other important contexts, but predominantly it is found in the form of single strands. Some of the bases are nevertheless paired, for there are regions of internal complementarity. Generally these are close together in the sequence, so that the chain can achieve the maximum amount of base-pairing by forming hairpins. But in between these double helices the chain forms a floppy coil:

Figure 19 RNA: the straight regions are lengths of a double helix.

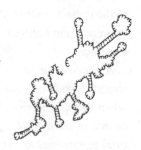

RNA molecules serve several functions in the cell, especially in the synthesis of proteins. First there is messenger RNA (mRNA), which contains a gene sequence, copied from the DNA with the aid of an enzyme, when the cell requires the protein specified by that gene to be manufactured. The mRNA acts as a punched tape, fed into a reading head to be scanned. The reading head is called a ribosome, a structure composed of proteins and three different molecules of RNA, collectively known as ribosomal RNA (rRNA). The tape is held taut and the correct amino acids corresponding to the code-word, or *codon*,* of three successive nucleotides in the sequence is conveyed to the site. The vehicle for the amino acid is the third form of RNA, called transfer RNA (tRNA). There are transfer RNAs for all of the twenty amino acids. The tRNA molecules are loaded with their cognate amino acid by an enzyme (a different one for each amino acid). At its other end the RNA has a loop containing a triplet of nucleotides complementary to the coding triplet poised for action on the ribosome. So for instance if that triplet happens to be CGA (the code word for the amino acid arginine), the arginine-bearing tRNA must have a loop containing the complementary triplet, GCU. The apposing triplets on mRNA and tRNA bind to each other, and the arginine is transferred by another enzyme from the tRNA to the growing protein chain. The mRNA tape moves on a notch in preparation for the arrival of the next amino acid in the protein sequence.

A remarkable form of RNA is the *ribozyme*, a single-strand structure with the capacity—highly improbable as it appeared when first discovered—to catalyse the fracture of its own chain or that of other RNAs. That something so restricted in its structural possibilities as a nucleic acid could function as an enzyme

---

* With four different kinds of base there are sixty-four possible triplets. One of these specifies the signal to start translating a gene, and another comes at the end and is the signal to stop. All the other triplets code for an amino acid. Their identity is the famous *genetic code*. The sequence of nucleotides in the DNA of all the chromosomes defines the *genome*.

remains surprising, but it is true. Our cells also contain many small RNAs, which act to suppress selected genes by virtue of complementary sequences to those in the messenger RNA. These are called silencer RNAs or siRNA, and they give promise in the therapy of a variety of diseases.

The DNA computer

Contemplating the nature of the genetic code and the processes of biological information transfer, several mathematicians and computer scientists were smitten with a wild surmise. It is the enormous density of information stored in a strand of genomic DNA which seized their attention. One of the leading proselytizers for the exploitation of this capacity for computing, L. M. Adleman, has calculated that 1 g of DNA, occupying, when dry, a volume of 1 cubic centimetre, has a storage capacity equivalent to 10^{12} (one trillion) compact discs. But to manipulate information stored in this way in a DNA computer requires imagination and a great deal of biochemistry, which we will leave for later.

5

THE PROTEAN ELEMENT: CARBON IN NEW GUISES

Whether anything made of only the atoms of a single element can decently be classed as a polymer is a question fit for pedants. But new forms of this most remarkable of elements, discovered in the past two decades or so, have opened new vistas to the eager gaze of chemists and physicists. They are, moreover, finding practical applications of a kind which would have seemed closer to science fiction than science only a few years ago. So all this excitement can scarcely be disregarded here.

Graphene and the buckyball

Until late in the past century only two ordered forms of carbon were known to science. There was diamond, with its perfect crystal structure, and there was graphite, also crystalline, but black and flaky and not at all transparent. Besides those there are only coal, coke, soot, lampblack, and the many kinds of charcoal. The graphite structure reflects its properties. It is made up of sheets of carbon atoms arranged in a hexagonal lattice, like fused benzene rings (p. 17), looking like a honeycomb, and with weak bonds between adjacent sheets. This means that the graphite easily forms flakes when the sheets split away from each other, and, further, because the sheets can slide over each other, that graphite finds use as a lubricant. It is also a good conductor of electricity. Quite recently, in 2004, researchers in Manchester found that they could mechanically peel single sheets—two-dimensional crystals—from three-dimensional graphite crystals. (This can be done with just a piece of sticky tape.) They called the result *graphene*.

Figure 20 The honeycomb lattice of a graphene sheet, showing carbon atoms and the bonds between them.

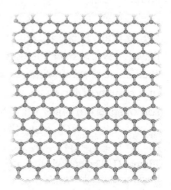

These sheets are semiconductors—they are intermediate in their electrical character between conductors and insulators. Transistors are made of such materials, and applications of this sort for graphenes are anticipated. But graphene sheets are, as we shall see later, semiconductors of a kind not previously encountered, with properties that may one day change the face of electronics.

A wholly new form of carbon came to light in 1985 and caused an immediate sensation. Its discovery was an accident, foreshadowed by two workers in Texas, Robert Curl and Richard Smalley, who detected a species of carbon in outer space, with a size implying that it was made up of about 60 atoms. Harry Kroto at the University of Sussex was interested at this time in the nature of unidentified matter in outer space. To simulate interstellar conditions he vaporized carbon from a graphite rod by blasting it with a laser. The operation was carried out in an atmosphere of helium, an inert gas, so that there was nothing with which carbon atoms could combine. What resulted were two products, of which one, with the composition C_{60}, predominated. Because the number of carbon atoms was so precisely defined, it could be assumed that the new species must have the form of a closed shell, rather than an open polymer, which could grow by the addition of more carbons and would therefore be expected to vary in size. Now the only way to generate a closed spherical shell made up of 60 identical atoms is to place them at the corners of a system of hexagons and pentagons. One cannot make this sort of spherical cage from hexagons alone, as in graphite—a fact

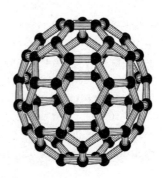

Figure 21 The buckyball: buckminsterfullerene, C_{60}. The 60 carbon atoms in the spherical molecule are shown by the black blobs, joined by chemical bonds.

known to the ancient Greek mathematicians, and to the makers of footballs:

Kroto and the two Texans who had made the discovery in outer space, Curl and Smalley, joined forces and determined the three-dimensional structure of C_{60}. The outcome could have been predicted: the cage has thirty-two faces, twenty of them hexagonal, the other twelve pentagonal. The strain involved in deforming the angles between the carbon atoms from the preferred hexagonal to the pentagonal geometry is small enough not to matter much, and thus the molecule is very stable. Other shell structures exist, but are less favourable, because they entail greater distortion of the carbon bonds, and are therefore created in much smaller amounts.

All three workers shared the Nobel Prize in 1996. But what name to give the singular new molecule—'the most beautiful molecule' as one of the discoverers called it? The structure immediately brought to mind the geodesic domes, constructed by the American architect and engineer Buckminster Fuller, and so the fanciful name buckminsterfullerene was coined, and has stuck, though 'buckyball' often now takes its place in speech and even print. The terms baffled the members of the House of Lords in 1991, when the question arose in debate as to 'what steps are being taken to encourage the use of buckminsterfullerene in science and industry?' A noble Baroness demanded to know whether 'this thing is animal, vegetable or mineral', and a government minister noted that Professor Kroto had described it 'as bearing the same relationship to a football as a football does to the Earth. In other words, he assured their Lordships, 'it is an extremely small molecule' (implying perhaps that, were it a large molecule, it might be the size of an actual football). Since its original discovery buckminsterfullerene has been found to occur in our midst, and not only in outer space—in the atmosphere after thunderstorms, in soot, and even in a mineral excavated in Russia.

The mighty nanotube

Various applications for buckminsterfullerene (and other fullerenes, as they are called, with cages of fewer carbon atoms) in lubrication, transistors, drug delivery, storage of hydrogen (because the carbon atoms can react with hydrogen without disturbing the cage structure), photovoltaic devices, and so on, have been envisaged. But in terms of practical interest, buckminsterfullerene has been largely supplanted by the *carbon nanotube*. (The name, like that of the nanometre, one-billionth of a metre, comes from the Greek *nanos*, a dwarf.) This is in effect a tube made from a graphene sheet. It is narrower than a human hair by a factor of up to about 100,000; it is thousands of times as long as it is wide, and its ends are capped by a fullerene-type dome:

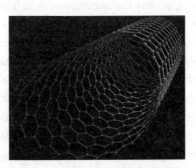

Figure 22 A carbon nanotube, viewed end-on. Note the hexagonal lattice of carbon atoms.

Carbon nanotubes were first described by two Russian physical chemists, L. V. Radushkevich and V. M. Lukyanovich, in 1952. Since their paper appeared in a Russian journal, there was scant chance that their Western confrères would have been able to read it or procure a translation, and even less that the journal would have reached their libraries, for at that time paranoia ruled, allowing little exchange of journals or information between the Soviet Union and the West. And so another quarter of a century passed before nanotubes were rediscovered, and it was only in 1991 that Iijima in Japan and Bethune at the IBM laboratories in the USA prepared nanotubes in sufficient quantity for detailed studies. The material was recovered from the dust formed when an arc discharge burned between two graphite rods. Other methods were later developed, especially deposition of carbon from a mixture of gases, made to react at a surface covered with minute granules of a catalyst; heating to a high temperature then converts the deposit into nanotubes, their size determined by that of the catalyst granules.

Since carbon nanotubes became objects of intensive interest, it has emerged that they are actually quite commonplace, for they are generated in ordinary coal gas or petrol flames; but to see them requires an electron microscope. Today one can buy nanotubes by the gram, though still at considerable cost. The proliferation of nanotube technology has elicited concerns about dangers to health. The nano-forms of carbon are often referred to by alarmists—journalists and politicians mainly—as the 'grey goo', a sinister new threat to man's survival, oozing out of the laboratories. There is indeed some evidence of effects on cells, perhaps on the lungs, and it has been reported that nanotubes will kill bacteria by puncturing the cell wall. So it has very properly been urged that caution should be exercised in the handling of these materials (just like asbestos and toxins in general). Nevertheless, groups of lay critics, such as the 'angels' of THRONG (The Heavenly Righteous Opposed to Nanotech Greed) in Britain, have been barricading laboratories and disrupting conferences to make their point.

Carbon nanotubes come in several forms. They may be single-walled—a single rolled graphene sheet, double- or multiwalled, with two or more cylinders nested like Russian dolls, or they may be 'parchment rolled'—made from a single graphene sheet, twisted like a rolled-up newspaper. Not only that, but single-walled nanotubes can be generated from the graphene sheet in three ways: the hexagonal, 'honeycomb' lattice may be curled with the lines of fused hexagons in the direction of the tube axis, or they may run around the circumference. These forms are referred to as 'armchair' and 'zigzag'. In the third form, the graphene lattice is curled in a less symmetric disposition, so that the line of hexagons twists along the surface in a helix. The variants are illustrated below:

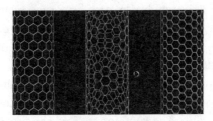

Figure 23 Three types of carbon nanotube: zigzag (left), chiral or twisted (centre), and armchair (right).

The helical form may be left- or right-handed (just like an α-helix in a protein (p. 35), but of course unlike protein or nucleic acid helices, it is not made of asymmetric units, so has no preference for one handedness over the other). This form is called 'chiral' from

101

the Greek *kheir*, a hand, for obvious reasons. The three symmetry types of nanotube differ in important ways in their properties.

Carbon nanotubes are truly extraordinary materials, the strongest and stiffest known to man, in terms of tensile strength and elastic modulus. (We will come to what exactly these terms imply later (p. 147).) Armchair nanotubes conduct electricity like metals, and, because of their enormous heat resistance, can tolerate current densities a thousand times greater than copper wire of the same diameter. The zigzag form, on the other hand, is a semiconductor (p. 96). Nanotubes are better conductors than metals of heat. We will have more to say later concerning the many uses, actual and potential, for carbon nanotubes. One pointer is their presence in the blades of swords made from the famous Damascus steel in the Middle East at the time of the Crusades. These swords were unparalleled in both their strength and the sharpness of their cutting edge. They were made from a special high-carbon steel, called *wootz*, developed in ancient times in India; there is evidence that the carbon came from wood and leaves added before smelting. Steels containing so much carbon (up to 2%) are brittle, and it is not known how the swordsmiths of the day overcame this problem, for the secrets of their methods were lost during the eighteenth century. It is known only that elaborate processes of hammering, heating, and annealing were involved, and that the steel also contained small amounts of other metals, such as vanadium (still used today to strengthen steel), manganese, and cobalt. Research in a German laboratory has revealed that the blades are traversed by multiwalled carbon nanotubes, observed by electron microscopy in the residue remaining when the metal is dissolved in acid. The conclusion is that the coveted properties of Damascus steel, including its characteristic wavelike pattern of bands, derived from the nanotubes in the steel.

Nanotubes in space

The colossal strength of carbon nanotubes, far beyond that of anything else ever created or seen, excepting only the brittle diamond, has revived a century-old spectre. The vision of the *space elevator* came to the Russian engineer and science fiction writer Konstantin Tsiolkovsky in 1895, provoked by the newly erected Eiffel tower in Paris. Tsiolkovsky was self-taught because his schooling more or less ceased when he became almost deaf at the age of ten. He is regarded—and not only in Russia—as the veritable father of space travel. His writings inspired the pioneers of rocketry, such as Werner von Braun. It was Tsiolkovsky who conceived the principles of multistage rockets, liquid fuel, space locks, and many more of the concepts on which the technology of space exploration is based.

In 1978 another visionary, Arthur C. Clarke, published a novel, *Fountains of Paradise*, in which the space elevator again featured, and the conceit is now often credited to him. With the advent of carbon nanotubes, it has been seriously taken up by NASA, the National Aeronautics and Space Administration, in the United States. The device is in essence a tethered cable extending from a point near the equator to a captive space station, moving in what is called a geosynchronous orbit (GSO)—one that rotates with the Earth. Before the carbon nanotube calculations showed that there were no known materials with the tensile strength, nor yet the stability and elasticity, to make the device remotely feasible. But now NASA engineers are working on the design. A tower some 50 km high will form the base of the elevator. The cable, made from nanotubes (which have the useful propensity to associate laterally with one another), would be held in place by a counterbalancing weight in space, rotating with the Earth. This might be a captive asteroid, on which a space colony will be established, and people and materials will be conveyed up the cable by an electromagnetic vehicle, like the trains that ply between the terminals in many airports, but lighter (thanks again to nanotube composite materials), and travelling at hundreds, perhaps thousands, of kilometres per hour. There seems to be no doubt in the NASA engineers' minds that the scheme is workable, nor that it will afford a vastly cheaper means of giving man access to space. When Arthur C. Clarke was asked his opinion of how long it might take to accomplish his vision, his reply was 'Probably 50 years after everyone has stopped laughing.' Clarke also formulated some rules governing the advance of technology. One of them stated that 'the only way of discovering the limits of the possible is to venture a little way past them into the impossible.' And another: 'Any sufficiently advanced technology is indistinguishable from magic.'

6

THE PLASTIC WORLD

Over the course of the nineteenth century some organic chemists produced unfriendly, insoluble, sticky substances by rather uncontrolled reactions, which we now know to have been polymers and the precursors of plastics. The first marketable material seems actually to have been derived from milk. The process is said to have been invented in about 1530 by a Bavarian Benedictine monk, by the name of Wolfgang Seidel, but he probably got the idea from earlier recipes. It was based on collecting the curd from sour milk, by filtration, and soaking it first in hot water and then in lye (an alkaline extract of wood ash). The result was a tough, sticky mass, which would have consisted largely of the protein casein. Seidel was searching for a substitute for cow horn, which was used, amongst other things, for making windows for lanterns. The horn had to be heated in steam and then flattened and shaped while hot. This was a demanding procedure, calling for much skill. The milk extract, by contrast, could be easily moulded while hot, and would keep its shape when cooled. Its durability was later improved by addition to the extracted mass of powdered porcelain or similar material. Colouring matter could also be introduced.

Some 400 years on, in 1897, a German organic chemist, Adolf Spitteler, tried a more scientific approach. He added to milk the simple chemical formaldehyde, which was known to react with substances containing the amino group, $-NH_2$, and introduce covalent bonds between them. And amino groups, of course, abound in proteins. The story goes that the credit for the experiment really belongs to the laboratory cat, which upset a vessel containing formaldehyde into her saucer of milk. In any event, a hard horny mass resulted, which Spitteler patented under the name of Galalith (from the Greek, meaning milk stone), and which could be coloured and moulded to make items of paste jewellery. This, it could be said, was the hesitant start of the plastics industry. Nor was it the last time that proteins made an appearance in that industry. Perhaps the most famous example was a car built by the Ford motor company in 1942, with bodywork of chemically treated soyabean and wheat protein— 'part salad and part automobile', according to one newspaper. It was indistinguishable in form from a standard model, but was a good deal lighter. When unveiled, it excited much mirth—the only car that could be either driven or salted and eaten. Henry Ford had hazarded that one would soon be able 'to grow most of an automobile'. The picture shows the ageing autocrat attacking the plastic bodywork with an axe to show that it would not dent or crack. But the war intervened and nothing more was heard of the edible car. Rather surprisingly, the Japanese motor companies have returned to this scheme, in the hope apparently of luring green-minded customers. So far, it seems, they have had limited success, and have produced only fittings, such as dashboards and seat covers, from bioplastics (as we must now call them) made from maize.

Figure 24 Henry Ford attacking the 'edible car'.

Rubber

On the other hand, materials that we now know to be polymeric were already in wide domestic and industrial use many centuries ago. A trivial example is the use, since prehistoric times, of egg-white, blood plasma, and gelatin from bones in varnishes (even on cave paintings) and sizes. Rubber, which was known to the Mayan Indians of Mexico as early as the eleventh century, is of altogether deeper interest. The Mayans collected and cured the latex from the rubber trees to make bouncing balls for ritual games. (The penalty for the losers was commonly decapitation, and the severed heads might then be converted, by means of a coating of latex, into balls for the next round of the competition.) The latex oozes from the rubber tree when the bark is cut, hence the European name, caoutchouc, from 'weeping wood' in the Mayan language. When dried it set to a solid mass, for which the Mayans and other Indian tribes found practical uses beyond ball games—in medicine and surgery, and as a coating for fabric to give protection against rain (although, as more recent European experience indicates, this could have had only modest success). The related gutta percha trees grew in Southeast Asia, and were tapped for their latex, which was used as bird lime, but apparently for little else.

Rubber, or 'elastic gum', came to Europe in the eighteenth century. Charles Marie de La Condamine had been despatched by the French Academy of Sciences to measure a part of the meridian passing through Peru. On his return in 1736 he reported that the local Indians were using the cured latex to make waterproof boots and pear-shaped bottles, which ejected jets of liquid ('like a syringe') when squeezed. The new substance excited much curiosity in Europe. It was the English savant, Joseph Priestley who discovered in about 1770 that blocks of the recovered latex could be used to rub out pencil marks (by scraping off the surface of the paper), whence the name India rubber for the material (since it was believed to come from the West Indies).

Then in 1824 a resourceful Scottish chemist, Charles Mackintosh, found a good solvent—naphtha, one of the oily fractions derived from coal tar—for the latex. He also of course used the solution to coat fabrics, and found that two sheets of cloth glued together with his rubber solution were indeed waterproof. The material was not altogether satisfactory, for it would become sticky in hot weather and it emitted a disagreeable smell. Meanwhile, Charles Goodyear in America had been grappling with the problem of how to prevent rubber from softening when warm, and hardening and cracking in the cold. After many years of unsuccessful and financially ruinous toil, he hit on the answer quite by chance. On heating it with sulphur (a sample of the mix having been accidentally left on a hot kitchen stove) the desired effect was achieved. This process introduces cross-links between the polymer chains (although this was not understood at the time) and was given the name vulcanization. When taken to the limit it produced a hard substance, dubbed ebonite (or, by some manufacturers, vulcanite).

Over the ensuing years, many chemists tried to discover the chemical structure of rubber. It could be broken down by high temperature, and distillation would then yield an oil, which in turn gave rise to a liquid with a low boiling point. It turned out to be a pure substance, and was given the name isoprene with the formula

$$\begin{matrix} CH_3 \\ \diagdown \\ \diagup \\ CH_2 \end{matrix} C - CH = CH_2$$

Sir William Tilden, Professor of Chemistry at what is now Imperial College in London, who had done the work, also tried, with some success, to turn isoprene back into rubber. The methods available at the time (around 1900) to measure molecular weights all indicated that rubber must consist of very large molecules (even if the zealots of the colloid school clung to their own interpretation). In 1931 the structure of the raw rubber was finally established:

$$\underset{|}{CH_3} \qquad\qquad \underset{|}{CH_3}$$
$$-CH_2-C=CH-CH_2-CH_2-C=CH-CH_2-$$

In the 1920s X-ray diffraction analysis emerged as the supreme method for determining the geometrical structure of molecules, but only in crystalline or semi-crystalline materials, in which the molecules formed a regular array. By this means it was discovered that when rubber was stretched it became ordered, and also that the chain was in the *cis* configuration. This is an important concept, which we shall encounter repeatedly in what follows. To reiterate (p. 16), whereas the atoms on either side of a single bond can rotate about that bond, such freedom is denied to the atoms attached to carbons joined by a double bond; so in the polyisoprene chain of rubber we might have either of two configurations; in one (the *cis* form) the methyl ($-CH_3$) groups are on the same side relative to the double bond, in the other (the *trans* form) they are on opposite sides. Now, while in rubber the *cis* structure prevails, the chemically identical polyisoprene chain that occurs in gutta percha, the latex of a tree native to Southeast Asia, is in the *trans* form. The rubber and gutta percha chains are therefore written like this:

cis
rubber

trans
gutta percha

The upshot is that their physical properties are quite different, for gutta percha is inelastic and hard, becoming rubbery only when heated. Yet, when the first sample was sent to England from Singapore in 1839, its virtues were soon recognized, especially as an insulator for electric cables.

Indeed by 1847 the Gutta-Percha Company was already producing insulated cables. One such was laid beneath the English Channel between Dover and Cap Gris Nez, but only a few telegraphic messages were sent before it failed. The same fate overtook the first transatlantic cable. In 1857, through the efforts of the visionary American entrepreneur, Cyrus Field, two ships set out, one from Ireland, the other from Newfoundland, each carrying 1200 tons of insulated cable (for a load of 2400 tons would have sunk any ship then available). The plan was to meet in the middle and splice the two halves, but after a few days the westward cable broke. The next year another attempt succeeded, amid much celebration. Queen Victoria sent a message of ninety-nine words to President Buchanan in Washington (which took $16^1/_2$ hours to transmit), but the jubilation was short-lived, for after only 732 messages had been sent the cable broke down irretrievably. An adviser to the project was the physicist William Thomson, later Lord Kelvin of Largs, who realized that the transmission voltage was too high and the gutta percha insulation inadequate. Thomson not only developed a low-voltage transmission system, but insisted that the gutta percha should be applied in several layers, so that inevitable defects in any one layer would be covered up by the next. The improved cable was laid in 1865, but again broke. It was recovered the following year, just after a second cable had been successfully laid, and so the Global Village was born. More mundane uses were discovered for gutta percha: it formed the core of the modern golf ball, and changed the character of the game, while dentists still use it to fill root canals.

111

Gutta percha, though, was small beer compared to rubber which, following Goodyear's serendipitous discovery, quickly found many new uses. But it was with the invention of the pneumatic tyre that demand abruptly grew. By then seeds of the South American rubber tree had been illicitly brought from Brazil to England by a planter, Henry Wickham. He managed to smuggle out 70,000 of these delicate seeds, a few of which were germinated in the tropical herbarium of Kew Gardens outside London. The year was 1876, and it was Wickham's act of piracy (for which he was rewarded by a knighthood) that eventually broke the monopoly of the South American, mainly Brazilian, rubber barons. These men had built up a profitable industry, which relied on savage exploitation of Indian labourers. Exploitation now began elsewhere, for the seedlings were sent to Ceylon (Sri Lanka) and thence to Singapore. Where there had been rainforests in Malaya and the Dutch East Indies, rubber plantations sprang up; these, and later others in Thailand, became the principal source of European and American rubber. The supply would be interrupted only when the forests were seized by the Japanese during World War II. Nearly all the world's natural rubber is still made from the latex of a single species of tree, *Hevea brasiliensis*.

In the first decade of the twentieth century, with the advent of the motor industry, rubber consumption doubled, the price rose vertiginously, and with it the incentive to produce a synthetic version. It was a Russian chemist, Ivan Kondatov, who brought off the first true synthesis of rubber in 1900, starting from the close relative of the natural monomer, methylisoprene. (The result was that the synthetic rubber industry came to flourish in Russia, and the Soviet Union remained the leading producer until 1940, when it was overtaken by the United States.) But the German chemists were not slow to follow where Kondatov had led, and in 1912 the

Kaiser's motor car took to the road on tyres of the new 'methyl rubber'. And yet the demand for rubber continued to grow and the market responded with ever higher prices, so the demand for cheaper, stronger, and more durable synthetic products became more insistent.

In 1931, as we shall see, the American company, DuPont announced the creation of *neoprene*, made by polymerizing chloroprene, a chlorine-containing derivative of isoprene:

$$\cdots-CH_2\overset{\displaystyle H}{\underset{\displaystyle}{\diagdown}}C=C\overset{\displaystyle CH_2-\cdots}{\underset{\displaystyle Cl}{\diagup}}$$

Improved methods of vulcanization, and the addition of fillers, most often carbon black powder, to dilute the rubber, reduced the price and improved the reliability of the product. The issue grew in importance with the coming of World War II. Rubber became scarce while the need increased. In Germany the situation was critical. A synthetic rubber, known as Buna (from the starting monomer, *bu*tadiene and the metallic sodium, in German *na*trium, used to initiate polymerization), had been produced on a laboratory scale, but it was expensive and its properties were far from ideal. A more complex polymer, with two mixed types of monomer unit, polymerized by improved procedures, gave better results. Some material was produced in the infamous Buna factory in Auschwitz. The preferred material for tyres proved later to be a polymer made from styrene and butadiene:

CH = CH$_2$

$$CH_2=CH-CH=CH_2$$

styrene   butadiene

the latter of which is simply isoprene missing its methyl group.

113

Rubbery elasticity

Why then are rubber and its synthetic analogues elastic? Rubbery elasticity is quite different in its origins from the stretching of a coiled spring. Vulcanized rubber is a highly cross-linked random-coil polymer, as pictured below on the left. When a stretching force is applied to it, the chains open out, as on the right:

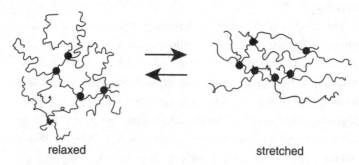

relaxed   stretched

Figure 25 Rubber relaxed and stretched.

It is the constraints on the freedom of the chain segments, imposed by the cross-links, which are the cause of the elasticity. Early in the nineteenth century an English chemist, John Gough, performed a simple and illuminating experiment. Apply, he said, a strip of rubber to your lips, stretch it sharply, and you will feel a sensation of warmth. Rubber emits heat when stretched, and conversely will contract, and not expand like most other materials, when heated.* So, since, as the First Law of Thermodynamics

---

* The link between the first and the second characteristic is encapsulated in Le Chatelier's Principle, formulated by the French chemist Henri Louis Le Chatelier in the nineteenth century. It is most simply stated like this: if to a system in equilibrium a constraint be applied, the equilibrium will shift in such a direction as to oppose the constraint. This is a great truth, which applies as much to economic and social disturbances as to chemical and physical equilibria. As a simple illustration,

tells us, work and heat are equivalent, the work done in stretching the rubber is greater than the resulting energy gained by the material, and the excess emerges as heat. How, in that case, is the extra work needed to counter the resistance to the stretch consumed? The answer was obvious to the redoubtable William Thomson, Lord Kelvin, and his friend James Prescott Joule,* two of the founders of thermodynamics: it goes into overcoming an increase in *entropy*.

The concept of entropy was defined by the nineteenth-century physicist Ludwig Boltzmann, who perceived that a system of particles will strive towards increased disorder. (Or, extending the principle to human affairs, 'things left to themselves go from bad to worse.') Boltzmann was a polymath, who lectured at Vienna University in physics, mathematics, and philosophy, and a man of legendary charm and humour. He was at the same time a depressive; and died in 1906 by his own hand, at the age of sixty-two; his gravestone bears the simple inscription, $S = k \ln W$, his famous equation expressing the relation between degree of order and entropy.

Entropy enters into every chemical and mechanical system. Put a spoonful of sugar in your coffee and the sucrose molecules will

consider what happens when a gas at equilibrium in a cylinder is heated: the only way in which it can oppose the change, in other words to cool, is to expand. And we know that when the gas in a refrigerator expands on passing through the expansion valve its temperature drops sharply.

* Joule was a north country man from a brewing dynasty. He inherited the brewery and ran it for much of his life. But his father had sent him as a boy to study chemistry with John Dalton (p. 2) in Manchester, and science became a predilection, which he pursued whenever he could. The unit of heat, the joule, is named after him, and he is remembered, among other things, for his many measurements of the relation between mechanical work and heat; in one of his early attempts he recorded the temperature of the water at the top and the bottom of a waterfall, for the energy lost in the descent could be calculated. He performed this experiment while hiking with his bride on their honeymoon.

quickly disperse themselves throughout the liquid; their gain in freedom to zoom around throughout the liquid volume amounts to an increase in the entropy of the coffee cup's contents. To get the sugar out again is a great deal harder, and requires a large (and precisely calculable) energy input to overcome the entropy of dispersion: a technological parallel is desalination. If you reduce the entropy of a system by restricting the freedom of its parts, you will need to pay in the coin of heat (or in modern vernacular, enthalpy). In the case of rubber, the polymer chains have a generous amount of freedom, limited by entanglement and by the cross-links, but when you stretch the rubber strip, the chains also stretch, more or less in the same direction. The freedom of the segments to take up a myriad of configurations is thereby greatly reduced. In other words their entropy is diminished. This is why the chains in rubber and like materials are referred to as 'entropy springs'. An important property of vulcanized rubber is, of course, its capacity to return to its original state when the stretching force is released, for the cross-links, randomly scattered throughout the polymer matrix, are immutable, and thus define the relaxed state. The entropy-spring concept describes the behaviour of a wide range of materials, natural and synthetic, known as *elastomers*.

The first true plastic

Cellulose fibres from plants, whether soft like cotton, flax, linen, and wool, or hard like hemp and jute, have served man since ancient times. Cotton and its relatives are in essence merely plant cells. This was apparent to a French savant, Anselm Payen, from examination under the microscope, and it was he who in 1839 coined the name cellulose (from cell and the terminal -ose,

116

denoting a sugar, since the relation to sugars was already known). Cellulose, together with several related polymers, also forms a large part of the cell-wall matter of wood, and is the essential constituent of paper. The structure of cellulose (p. 80) shows it to consist of sugar units linked in long chains. Chemists tried treating cellulose with a variety of reactive chemicals, and the first result of any consequence came in 1846, when Christian Schönbein, Professor of Chemistry at the University of Basel, hit by accident on the effect of the nitrating mixture of nitric and hydrochloric acids, known as *aqua regia* (nitration implying introduction of nitro groups, $-NO_2$, into a compound). The story went that he was working in the kitchen of his house, when he spilled some of the aqua regia, a ferociously powerful reagent (so-called since time immemorial on account of its ability to dissolve gold). To limit the damage to the work surface, and terrified of his wife's wrath, he hastily mopped up the puddle with a cotton apron, which he then rinsed with water and hung up to dry near the fire. A sudden smokeless conflagration ensued and the nitrocellulose industry came into being.

The first application, conceived by Schönbein, was gun cotton, which, indeed, he showed off at the Woolwich Arsenal in London; he even presented a brace of pheasants—the very first to fall to shot propelled by gun cotton—to Prince Albert. But a much more important discovery soon followed, namely that a less extensively nitrated cellulose, which became known as pyroxylin, could be dissolved in a variety of organic solvents (in which the parent cellulose was wholly insoluble). The viscous solution of pyroxylin, commonly in a mixed solvent of ethanol (drinking alcohol) and ether, was called collodion. This would dry when painted on a surface to form a tough colourless film. It quickly found its first serious application, in photography, as

117

the transparent matrix in which the photosensitive material was embedded. This 'wet-plate' process was introduced in 1851 and remained for the next twenty years the standard technique.

In 1862, a British chemist and inventor, Alexander Parkes, mixed pyroxyline (produced cheaply from cotton and paper waste) with camphor as plasticizer, and added solvents. This yielded a tough material which could be coloured in artistic ways and polished. He named it 'parkesine' and used it to manufacture many small articles. The new material aroused great interest, and new additives and improved methods of manufacture led to a more versatile substance, patented in the USA under the name celluloid. The consumption of camphor outran the supply, and laurel trees, from which it was derived, were planted in great numbers in Florida. But they did not thrive, so the material had to be imported at increasing cost. Nonetheless, celluloid was soon competing with rubber, especially for the manufacture of dental plates. But the big prize at the time was to contrive a replacement for the increasingly expensive ivory used to make billiard balls. A prize of $10,000—no mean sum at the time—had been put up by a manufacturing company for the first satisfactory ivory substitute. But despite strenuous research and extravagant claims of success, all efforts failed. For one thing, nitrocellulose was highly inflammable and explosive, and there were attested cases of alarming detonations when billiard balls collided at speed (once, according to a Colorado saloon-keeper, causing his clients to reach for their revolvers), and also of conflagrations. The green celluloid eye-shields, affected by newspapermen and others at the time, and dental plates could cause consternation, if not injury, on coming into contact with a lighted cigarette or cigar.

Celluloid film was responsible for a catastrophic fire during a film-show at a Paris charity bazaar in 1897, attended by a large section of Parisian high society. It ensued when the projectionist struck a match the better to see as he changed a reel in the dark. The wooden building was destroyed and 121 lives were lost, among them many *femmes de noblesse*. (It was said by the press that their mobility was impeded by their voluminous dresses and that they were trampled down by the menfolk heading for the exit.) The disaster provoked a torrent of sensational reportage, and swift legislation followed, prohibiting the use of nitrocellulose film. Many expensive explosions and fires also occurred in factories making celluloid products, but the material remained in vogue for detachable collars and cuffs, buttons, combs, and spectacle frames.

Another use for nitrocellulose was discovered by Hilaire Bernigaud, comte de Chardonnet, who in 1844 deposited at the *Académie des Sciences* in Paris a sealed recipe for the creation of fibres of the material, after first stripping some of its nitro groups by treatment with acid. The secret was revealed only in 1887, when the count was busy setting up a factory in Besançon to produce what came to be known as Chardonnet silk. This enjoyed a brief vogue until supplanted by the arrival of rayon.

Towards the end of the nineteenth century other cellulose derivatives ('substitutes for substitutes') were developed, of which the most important was cellulose acetate. (It took the place of nitrocellulose in French cinemas.) It was (and is) formed by treating cellulose with acetic acid:

Here Ac represents the acetyl group CH_3CO-

119

Cellulose acetate found a number of new applications; in particular, it could be spun into fibres, which, for a time, were the basis of the only commercially viable man-made fabrics. In solution it proved useful as a dope to reinforce, amongst other things, the fabric wings of early aeroplanes; and, sandwiched between two plates of glass, it formed the original safety glass for car windscreens. Soon there appeared another cellulose derivative, generated by the addition of sulphur-containing groups: rayon. Fibres could be spun from its solutions and formed into fabrics. These found a place in haute couture, at a time when shiny slinky dresses were in fashion. The material remains in use for kidney dialysis filters and wound dressings.

And then, in 1908, a remarkable new form of cellulose was developed by a Swiss chemist, Jacques Brandenberger, who conceived the idea of a stain-repellent fabric while watching red wine soak into a tablecloth in a restaurant. His efforts to coat linen with a cellulose film resulted only in a brittle sheet, but then he found that the coating could be peeled away from the cloth as a thin flexible transparent film. This became cellophane, first used as a humidor wrapping for cigars. The merits of its impermeability to moisture and contaminants soon became apparent, and it instantly made a huge impact on the food industry. The miraculous nature (as it appeared) of the material was even celebrated in one of Cole Porter's lyrics:

> You're the purple light
> Of a summer night in Spain.
> You're the National Gallery,
> You're Garbo's salary,
> You're cellophane!

The Bakelite age

By the turn of the twentieth century there was a new glint in the eyes of industrial chemists: why not design fully synthetic polymers with properties of choice? There were many failed attempts before the first such material was developed in America by Leo Hendrik Baekeland, a Belgian chemist from Ghent who had made his fortune in America by selling his process for an improved photographic paper to George Eastman. Baekeland's parents, a cobbler and a housemaid, were illiterate, but evidently appreciated the advantages of an education and sent their son to the best school where he thrived, showing a particular bent for chemistry. At the University of Ghent, which had a strong chemistry department, Baekeland's talent was recognized, and at an unusually young age, he was offered a faculty position. But he chose, like many European scientists at the time, to do a tour of some of the notable centres of learning. Then in 1890 he and his new wife landed in New York, where, instead of pursuing his academic career, he took employment with a company producing photographic materials. The encounter with Eastman followed, and with his newfound wealth, Baekeland set himself up with a laboratory in a large mansion in the town of Yonkers, close to New York.

There, starting in 1902, he embarked on a study of the reaction between phenol (p. 16) and formaldehyde, $H_2C=O$, which the doyen of German organic chemistry, Adolf von Baeyer, had found some thirty years previously to generate a refractory tarry mass. Baekeland examined the effects of high temperatures and pressures on the outcome of the reaction, for which he constructed a closed vessel, called a bakelizer. After five years of meticulous research, Baekeland had a hard resin, which he intended initially as a substitute for shellac, a brittle natural resin used to make gramophone records. It was insoluble in all common solvents, and (unlike the cellulose-based materials) did not soften when heated. It was also found to have excellent electrical insulating properties. It was of course *Bakelite*. Baekeland prepared many variants of his resin in searching for powders that could be moulded, coloured, and adapted for particular purposes. All were highly cross-linked chains, such as Bakelite C:

Bakelite caused a sensation when unveiled at a meeting of the American Chemical Society in 1909. It found military applications during World War I (in shell casings and aeroplane propellers), and in the following decade it conquered the domestic market. An early revelation was that its elasticity resembled that of ivory, and so here at last was the solution to the billiard ball problem. During the high noon of art deco, designers of the day seized on the new material. Bakelite supplanted natural substances in everything from fountain pens, buttons, pipe stems, knitting needles, and ashtrays to telephones, washing machines, and all kinds of electric appliances. 'God said, "let Bakelite be", and all was plastic,' ran the saying. Less visible applications were laminates and varnishes. (In England, as early as 1905 James

Swinburne had marketed a lacquer called Damard, because it was 'damn hard', which started the country's phenolic resin industry.) The author John Mumford was engaged to write an encomium to Bakelite (*The Story of Bakelite*), published with much éclat in 1924.

Bakelite held undisputed sway for half a century, until the advent of urea-formaldehyde resins. A stimulus to the development of these materials was the desire for brighter colours for household objects. Bakelite (phenolic) resins emerged in naturally sombre hues, and would therefore accept only darker colours. Resins made from urea and thiourea

$$O = C \begin{smallmatrix} NH_2 \\ \\ NH_2 \end{smallmatrix} \qquad\qquad S = C \begin{smallmatrix} NH_2 \\ \\ NH_2 \end{smallmatrix}$$

by contrast, were almost colourless and translucent. Mixed with a white filler, such as cellulose powder, they could be made white, cream, or any colour, pale or bright, or given a marble or alabaster texture. Other compounds containing amino groups were also used as the basis of plastics. Among the most successful were resins made from formaldehyde and melamine, a ring compound, containing three amino groups. These materials were more expensive than Bakelite or urea resins, but had improved heat resistance, and a pleasing porcelain-like texture. Many Bakelite, urea, and melamine resin articles are now collectors' pieces, displayed in design museums around the world.

The road to nylon: the age of Carothers

In 1927 there appeared on the scene the most remarkable figure in the history of polymer science. Wallace Hume Carothers was born in rural Iowa in 1896. He was a high school prodigy, excelling in all subjects, but the best his penurious father could offer him in the way of further education was a place at Commercial College in Des Moines, where he himself taught. With a qualification in accountancy, the young Wallace found a teaching position at an insignificant college in a small town in Missouri. There in his spare time he took a degree in chemistry in only two years. His talents were recognized and he secured a place at the University of Illinois to work under one of the pre-eminent organic chemists of the time, Roger Adams. Adams developed an affection and high regard for Carothers, and found him a faculty teaching position. After two years of this Carothers was recruited by James Conant, a distinguished chemist (and later College President) at Harvard. An illustrious academic career beckoned, but Carothers did not enjoy teaching, so when he received an offer from the mighty DuPont Company to head their new basic (pure) science laboratory, 'Purity Hall', just constructed in Wilmington, Delaware, he accepted, with little soul-searching, and to Conant's regret. He was thirty-one years old.

The DuPont management conceived the aim of the laboratory as the pursuit of basic research in broadly defined areas, one of which was to be polymers. This suited Carothers well enough, and it was not long before he had shaped the future of polymer chemistry. He had no patience with the benighted colloid chemists' noisy insistence that very large molecules did not exist, nor did he think much of their principal adversary, Hermann Staudinger, whose chemical syntheses he regarded as poorly controlled. His own methods were rigorous and precise. He defined the types of reaction that led to the formation of polymers, and developed *copolymers*, containing more than one kind of monomer unit. His first important achievement, which came in 1930, was the synthesis of neoprene (p. 113). It was followed in the same month by the first polyester. An ester is a compound formed from an acid, R—COOH, say, and an alcohol, R'—OH. Under the right conditions these will react, yielding the compound

$$R-C\overset{\displaystyle O}{\underset{\displaystyle O-OR'}{\big\|}}$$

where, as before R and R' are two different groups. So, for example if R is a methyl group, —CH$_3$, which means that the acid is acetic acid, and R' is an ethyl group, so that the alcohol is ethyl alcohol, as in whisky and wine, the ester

$$CH_3-C\overset{\displaystyle O}{\underset{\displaystyle O-C_2H_5}{\big\|}}$$

is ethyl acetate (a good solvent, with a characteristic smell of boiled sweets).

Carothers's method was to use a pair of bifunctional reagents, that is to say, an acid twice over, possessing two —COOH groups (a dicarboxylic acid), and similarly an alcohol with two —OH groups. We then have HOOC—R—COOH and HO—R'—OH, which can be expected to react at both ends, and thus form a long chain: ... RCOOR'OOCRCOOR'COO.... (Both Carothers's R and R' contained more carbons than the methyl and ethyl groups shown, for illustration, above.) Carothers, moreover, found ingenious ways of achieving chains much longer than had been seen previously. From the resulting mass of polymer strong, lustrous, elastic filaments could be drawn, and the vision of a synthetic silk appeared before the eyes of the enchanted DuPont executives. But the starting materials were expensive, and the melting

125

temperature uncomfortably low. Improved polymers followed, but the real breakthrough came in 1935 with the synthesis of a polyamide—a polymer, that is to say, with a backbone like a protein. Starting from a diamine (a compound of the type $H_2N-R-NH_2$) and a dicarboxylic acid, each with six carbon atoms, Carothers and his assistants produced the so-called polymer 6-6, later renamed Polymer 66 (one of eighty-one polyamides which they synthesized and analysed). Fibres drawn from this substance were stronger than any natural material then known, and had remarkable stability, with a melting point of 263 °C. X-ray crystallographic analysis revealed that the polymer molecules making up such fibres were highly ordered, all of them aligned in the direction of pulling. The fibres were resistant to heat, and were elastic and of a tensile strength far greater than that of cotton or silk. They could be turned into toothbrush bristles, and they could be spun and woven into a sheer, washable shrink-proof fabric.

After long deliberation, the name selected (out of several hundred suggestions) for this numinous material was nylon. (A journalistic canard had it that it was an acronym of Now You Lousy Old Nipponese, for the textile community was already gloating over the forthcoming ruin of the Japanese silk industry.) The press portrayed nylon as an indestructible fabric, resistant to laddering, and stronger than steel. The public brouhaha, which the announcement of the new wonder fabric elicited, is hard now to grasp. The problems of large-scale production took some years to solve, but in 1939 the first 4 million pairs of nylon stockings released to the fashion houses of New York were sold in hours to rampaging hordes of women and their male proxies. There were minor vicissitudes, as when an enterprising journalist examined the DuPont patent application and noticed that the diamide used in synthesizing the nylon polymer was cadaverine (in which R in the formula above is C_5H_{10}). As its name implies, this vile-smelling substance is generated in putrefying animal matter, and when that was reported in the press, the rumour quickly spread that nylon was derived from human corpses. The story took a year or two to die, countered by a proclamation by DuPont that nylon was made from 'coal, air and water', which was more or less true, since the precursors were indeed coal tar products.

Thereafter, with war impending, the entire output of nylon was diverted into the manufacture of parachutes, and in fact when the United States finally entered the war, there were appeals to American women to sacrifice their nylons in the patriotic interest, to be melted down for parachute fabric. A popular song in 1943 went 'When the nylons bloom again'. It was not until 1945 that the ardent public demand for nylon stockings could be gratified, and when in September of that year the first specimens reached the market, the 'nylon riots' began, and for some months shops were besieged and women trampled underfoot. And what of the begetter of the nylon age? Wallace Carothers did not live to see the fruits of what he had wrought. He was a depressive, prone to spells of corrosive self-doubt and lassitude, and in such a state, at the age of forty-one, he locked himself in a hotel room in Philadelphia and drank cyanide.

The heirs of Carothers

By the time Carothers died, many industrial chemists had read the signs and were busy synthesizing new artificial materials. In 1933 chemists at ICI in England left a reaction mixture of acetylene gas ($HC{\equiv}CH$) and a catalyst in a high-pressure vessel over a weekend and found on Monday morning that the pressure had dropped and instead of gas the vessel contained a white powdery solid. This was polyethylene (p. 14), or *polythene*, which proved to be an exceptionally valuable polymer. It was resistant to breakdown by electric current, and could be used to isolate electrodes from each other. It was an essential element of airborne radar equipment in World War II. Later Earl Tupper in America moulded food containers from polythene. Vinyl plastics, such as polyvinyl chloride (PVC)

$$\cdots - CH_2 - \underset{\underset{Cl}{|}}{CH} - CH_2 - \underset{\underset{Cl}{|}}{CH} - CH_2 - \underset{\underset{Cl}{|}}{CH} - \cdots$$

and polyvinyl acetate

$$\cdots - CH_2 - \underset{\underset{O=C-CH_3}{|}}{CH} - CH_2 - \underset{\underset{O=C-CH_3}{|}}{CH} - CH_2 - \underset{\underset{O=C-CH_3}{|}}{CH} - \cdots$$

came on stream, and also the related polystyrene,

$$\cdots - CH_2 - CH - CH_2 - CH - CH_2 - CH - \cdots$$

made from styrene, which is vinylbenzene (p. 113).

Foam, made by blowing gas through the polymer melt under pressure, came into use as a heat insulator, in houses and coffee cups. In 1930 the Union Carbide Company had already unveiled a new vinyl plastic called Vinylite, a copolymer of polyvinyl chloride and polyvinyl acetate. It had little to commend it, compared to what had gone before, but at the grandiosely named *Century of Progress Exposition* in Chicago in 1933, the Vinyl House, made

Figure 26 The nylon riots in 1945.

entirely of Vinylite, from the walls and floors to the furniture and even the towels, excited much attention—'the house that chemists built'.

At about the same time Otto Röhm, a German American, made the first acrylic resin, polymethylmethacrylate, known as Plexiglas or Perspex:

$$\cdots - CH_2 - \underset{\underset{O=C-O-CH_3}{|}}{\overset{\overset{CH_3}{|}}{C}} - CH_2 - \underset{\underset{O=C-O-CH_3}{|}}{\overset{\overset{CH_3}{|}}{C}} - CH_2 - \underset{\underset{O=C-O-CH_3}{|}}{\overset{\overset{CH_3}{|}}{C}} - \cdots$$

This transparent plastic attained incalculable importance in the war for the construction of aircraft parts, such as windows and cockpit canopies. After the war, the manufacture of jukebox consumed the largest amount of Plexiglas. Röhm had also used a solution of the same material, which he called Plexigum, to unite two sheets of glass. The new polymer supplanted (because it was durable and did not turn yellow) nitrocellulose for the manufacture of safety glass.

The virtues of Saran wrap (cling film) were discerned by the Dow Chemical Company, which originally developed it as a covering for furniture with the special advantage that it could be cleaned with industrial solvents. It emerged from an accidental observation by a laboratory assistant, who, while washing up the day's dirty glassware, found an interesting sticky insoluble mass in a reaction vessel. It was polyvinylidene chloride, which is PVC with an extra chlorine atom in each monomer unit:

$$-CH_2-CCl_2-CH_2-CCl_2-CH_2-$$

One of its first uses was as a spray coating for fighter aircraft during the war to protect them against corrosion. Its use as a food wrapping (no more than a hundredth of a millimetre thick) came later.

In 1938 yet another serendipitous observation led to the discovery of polytetrafluoroethylene, PTFE or Teflon, as it became known. In this polymer atoms of fluorine replace the hydrogen of polyethylene: ... $-CF_2-CF_2-CF_2-$ .... Roy Plunkett, a chemist at DuPont, was searching for a better refrigerant, using as a reagent the gas tetrafluoroethylene ($F_2C=CF_2$). When he opened the valve on a cylinder containing the gas one day, nothing came out. Instead of discarding the cylinder, Plunkett weighed it; the weight indicated that it should have been full of gas. There had been no suggestion that tetrafluoroethylene was capable of spontaneous polymerization, but when the cylinder was sawed open it was found to contain a white waxy substance, resembling polyethylene. Plunkett guessed that it would have remarkable qualities, for fluorine is a highly reactive element, which tends to form exceptionally stable compounds. Such a one PTFE turned out to be: it was found to have unprecedented chemical resistance, did not decompose even at temperatures well above 300 °C, was more slippery (thus friction-free) than any known material, and had the valuable property of not wetting; so drops of water on a Teflon surface will not spread. It found its first application in the Manhattan Project. The gas diffusion process for preparing the fissile uranium isotope, U^{235}, for the first atom bomb had run into serious trouble, for the gas in question, uranium hexafluoride, was so reactive that it quickly destroyed all apparatus exposed to it. A Teflon coating solved the problem. The innumerable commercial and domestic uses of Teflon followed only from about 1948. From its modest beginnings in stain-repellent coatings it spread into the aerospace, building, communications, and transport industries, and of course into the kitchen.

We return finally to polyesters, which Carothers had first explored in his early years at DuPont. Although they were over-shadowed by the successes of the polyamides, research continued in the DuPont laboratories and elsewhere. Polyethylene tereph-thalate (PET), with the structure

$$\cdots -CH_2-CH_2-O-\underset{O}{\overset{\parallel}{C}}-\!\!\bigcirc\!\!-\underset{O}{\overset{\parallel}{C}}-O-CH_2-CH_2-O-\underset{O}{\overset{\parallel}{C}}-\!\!\bigcirc\!\!-\underset{O}{\overset{\parallel}{C}}-O-CH_2-CH_2-O-\underset{O}{\overset{\parallel}{C}}-\!\!\bigcirc\!\!-\underset{O}{\overset{\parallel}{C}}-\cdots$$

proved to be a versatile polymer, which gave rise to Mylar film and more rigid plastics used for bottles. But then in 1948, it also led to the first polyester fabric, called Terylene, which emerged from the laboratories of the UK Calico Printers Association, and almost at the same time from DuPont, who christened it Dacron. The fibre could be worked to resemble wool, and suits made from it could be washed instead of dry-cleaned. Many mixtures with natural wool or cotton were soon produced, each with its own advantages and disadvantages. In time chemically related polyesters, such as polypropylene terephthalate were developed for particular purposes, and there are now too many to enumerate.

Even DuPont, however, encountered the odd banana skin on their triumphal path. Corfam, their artificial leather, proved a $70 million fiasco. It was nearly a decade in the making, and was meant for shoes that would not only 'breathe' but would resist wear and retain a high polish for ever. The material comprised a web of polyester fibres, impregnated with polyurethane and covered with a polyurethane coating. The soles were porous (a million pores per square centimetre, the advertising proclaimed). But the material was not compliant, feet felt hot, and the appearance of the product did not please buyers when it came on the market in 1963. DuPont eventually sold the patent for a derisory sum to an organization in Poland, and thence into oblivion.

THE QUIDDITY OF POLYMERS: SHAPES, SIZES, AND THEIR EFFECTS

The twisted backbones

It is time now to consider what form the various polymer molecules take, and how a grasp of the science has led to the development of a dazzling range of extraordinary materials.

We have already seen that the atoms at either end of a carbon–carbon double bond are more or less locked in place in either the *cis* or the *trans* configuration, while a single bond allows free rotation of those atoms about its axis. This much is true, but the dispositions of the atoms along the polymer chain is nevertheless limited by the nature of a covalent bond, which specifies favourable (i.e. low-energy) positions of the angle of rotation, also called the *torsional angle*. Free rotation means only that the atoms can flip from one favourable torsional state to another with some ease. The diagram will make clear the combinations of torsional angles available in a substituted ethane molecule, HRC–CRH, where R might be for example a methyl group, $-CH_3$, or perhaps a chlorine or the much larger bromine atom. Bear in mind that the four valence bonds of each carbon point to the corners of a tetrahedron, or pyramid (p. 19). There is a symmetry to the favoured torsional angles: the two central carbon atoms can take any of three positions relative to each other, related by twists of $120°$, one-third of a full rotation. If the R in the formula were simply hydrogen, we would have unsubstituted ethane, C_2H_6, or H_3C-CH_3, and the possible configurations reduce to two. Viewed end on, the sets of three hydrogen atoms on the two carbons can be in the eclipsed (*cis*) or staggered (*trans*) position relative to each other, and in perspective the molecule will look like this:

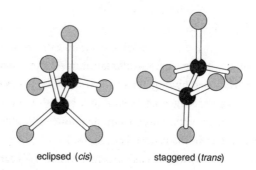

eclipsed (*cis*)   staggered (*trans*)

Figure 27 Stereoisomers of ethylene: *cis* and *trans*.

(bearing in mind that if the C−C bond points to one corner of a tetrahedron, the three hydrogens at the front will project out of the plane of the paper towards you, pointing to the other three corners, while those attached to the rear carbon will project backwards out of the plane away from you).

But suppose now that in place of one of the hydrogens on each carbon there is a different atom or group, R. In this case, there are two staggered states, according to whether the two R groups are related to the eclipsed state by a clockwise or an anticlockwise twist. Considerations of size now enter into the story, for the stick-and-ball representation of the molecule diverges from physical reality, in that a substituent like a methyl group is large compared to the length of the carbon–carbon bond, and so two such groups apposed at the ends of a single bond barely have room to coexist. Atoms, it should be recognized, are not hard objects like billiard balls, but consist of a cloud of electronic charge surrounding a nucleus. When two atoms approach too closely the negatively charged electron clouds repel each other, and create an energetically unfavourable situation, so that such configurations are avoided. If the substituents are large enough, the resulting *steric hindrance*—an insufficiency of space to accommodate them both—precludes the configuration altogether. Even in the eclipsed form of ethane, above, there is a small unfavourable interaction between the apposed hydrogen atoms, which makes it marginally less stable than the staggered (meaning that, although free rotation prevails, there will at any instant be a lot more molecules in the staggered than in the eclipsed configuration).

Let's persist now with the case in which the two Rs are methyl groups, making the molecule butane, C_4H_{10}. The diagram shows, with the methyl groups, R, represented by a black sphere (drawn unrealistically small for clarity) and the hydrogen atoms by a white sphere, that the molecule can, in principle, switch between any of six configurational positions (generated as before by twisting the rear carbon in increments of 60°). Two to these, with the methyl groups eclipsed, and with two hydrogens eclipsed, are unfavourable, the former grossly so, on account of close proximity of atoms. This leaves three more favourable states, as depicted below. The middle one, termed *staggered* or *anti*, with the methyl groups furthest apart, is the best. The two states on either side, attained by a 120° twist, are termed *gauche*, and depending on whether the twist is anticlockwise or clockwise are designated $g(-)$ and $g(+)$.

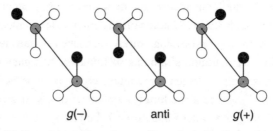

$g(-)$ anti $g(+)$

Figure 28 Rotation about carbon–carbon single bond: *gauche* and *anti*.

The difference between the two gauche forms, however, is illusory, because viewed from behind $g(-)$ turns into $g(+)$ and vice versa (which is why they are not called *gauche* and *droit*). Where this matters is in a longer chain, because then clearly two successive $g(-)$ bonds are not equivalent to a $g(-)$ followed by a $g(+)$, which will leave the ends pointing in quite different directions. The *anti* configuration remains the most highly populated because it is the most energetically favoured.

Consider now what this implies for a really long polyethylene chain with, say, 50,000 carbon atoms. If each successive pair of carbon atoms has a choice of three configurations, then the total number of possible chain configurations would be $3 \times 3 \times 3 \times 3 \ldots \times 3$, reaching $3^{50,000}$, a number too monstrous to contemplate. It means that at any instant (allowing for the great frequency—thousands of millions of times each second—of free rotations) no two molecules in a solution of polyethylene would have precisely the same configuration. The huge number, it is true, is attenuated by the relative prevalence of the *anti* state along the chain, and also by the exclusion of such combinations of torsional angles as would bring any two atoms in different parts of the chain into coincidence, for they cannot both occupy the same position in space. (The chain, in other words, engages in what is called a self-avoiding walk.) Nevertheless, the number of possible configurations remains so immense that each chain flickers constantly between a myriad of configurations. A chain of this type is called a random (or statistical or Gaussian) coil. Under the electron microscope even hugely long chains of this

Figure 29 DNA spilling out of a ruptured bacteriophage.

nature appear as tiny balls. Many polymers—if for example their chains contain double bonds, or especially if they have an internal structure like the double helix of DNA or the α-helix of proteins— are much stiffer. In such cases different theoretical descriptions are needed, and have been developed. DNA, then, looks distinctly elongated under the electron microscope, though by no means a stiff rod. Above is a picture of DNA spilling out of a ruptured bacteriophage (a virus that preys on bacteria), showing, incidentally, how tightly it must be packed within the volume of the blob in the centre.

Measuring sizes and shapes

Books have been written about the measurement of molecular weights of polymers, but a brief summary of the standard methods will suffice here (for there are many others). The first point to be made is that such measurements are really only possible on polymers in solution, and generally in dilute solution at that (though we can make an exception for a sufficiently rigid molecule of known structure like DNA, which can be measured up in electron microscope pictures like that above). Secondly, it is only biological polymers (and not all of those) which have a unique molecular weight. Man-made polymers are characterized by a distribution of sizes, most commonly Gaussian. A Gaussian distribution (defined early in the nineteenth century by the illustrious German mathematician Carl Friedrich Gauss) is depicted by the well-known bell-shaped curve, which describes such randomly distributed features as the heights or life expectancies of people in a population.

The first and simplest method developed for measuring molecular weights was end-group analysis. In a protein, for instance, the polypeptide chain (p. 83) may be expected to have a solitary NH_2 group at one end and a COOH group at the other. Or a synthetic polyester molecule (p. 125) will have an OH group at one end and (generally) a COOH group again at the other. If a chemical analysis is available to quantify the one or the other such group, it will deliver a molecular weight. But if there are, say, 10,000 monomer units in the chain there will be very few end-groups in each gram of polymer (as is most often the case), and the method becomes wildly inaccurate. A more useful way is by measuring the so-called colligative properties, which depend only on the number, and not the size or nature, of the molecules in solution. The most practical of these for large molecules is osmotic pressure, which depends on the inability of the molecules in question to penetrate a special (semi-permeable) membrane, with pores that allow free passage of the small solvent molecules. The dissolved polymer molecules will then exert a pressure on one side of the membrane, just like a gas inside a balloon. This can be measured, but again its sensitivity is poor when the molecules to be studied are large, for then, as before, each gram of substance will contain few molecules, compared to an equal weight of something small, like sugar or salt for instance.

The method on which Staudinger (p. 11) based practically all his conclusions about the nature of polymers was the measurement of viscosity of their solutions. He clung obdurately to the erroneous view that polymers were straight, rigid rods, and that there was a simple relation between viscosity and length. In actuality, the viscosity is determined not only by the concentration of the polymer and by its size, but also by its shape and the amount of solvent associated with it, for in general a polymer molecule attracts a certain number of solvent molecules. So while the viscosity will signal the presence of high-molecular-weight material, it is not so easily interpreted in quantitative terms. These uncertainties in his method provoked acrimony between the pugnacious Staudinger and his critics. An Austrian chemist, Herman Mark, who became a leader in the study of polymers, tried to convince Staudinger that his polymers were coils not rods and was denounced as a 'neo-micellist'—clearly the greatest insult that Staudinger could summon from his rich vocabulary—and the two men were reconciled only after Staudinger had mellowed in old age.

A far more solidly based method of size determination is analysis of *sedimentation* in the ultracentrifuge. This instrument spins the solution at speeds of up to around 60,000 revolutions per minute, and the heavy molecules, subjected to centrifugal force, move away from the centre of rotation. An optical system allows observation of the rate of sedimentation and also of the equilibrium that the distribution of molecules eventually reaches when the number of molecules driven outwards by the centrifugal force is balanced by the number diffusing back from a higher towards a lower concentration (just as the concentration of atmospheric gases is highest close to the earth, the gravitational pull balancing the tendency for the gases to diffuse away into the outer void). From this a molecular weight can be derived. For overcoming the formidable problems of constructing a turbine-driven ultracentrifuge with its optical system, and solving the theoretical problems associated with the method, the Swedish chemist, Theodor Svedberg received the Nobel Prize in 1926. Another classical method is to follow the rate at which the molecules diffuse from a region of high to one of low concentration. The larger the molecule (depending also on its shape), the slower its rate of movement. Finally large molecules in a solution will scatter light falling on it; if the molecules are very large and their concentration is high enough, a solution will actually look hazy to the naked eye. The phenomenon resembles the creation of fog by suspended water droplets. From the intensity of the scattered light the size of the molecules can be extracted, and from the manner in which it relates to the angle from which it is observed, a shape can in favourable circumstances be inferred. In some situations, better information can be obtained by the scattering of X-rays or of a beam of neutrons.

The most recent method of molecular weight measurement, now widely used, requires an expensive instrument, a mass spectrometer. A great advantage is that only a minute amount of sample is needed, and the accuracy for all but very large molecules surpasses that of other methods. The invention of the technique is generally attributed to the Cambridge physicist F. W. Aston, whose purpose was to identify isotopes (p. 2, f.n). He received the Nobel Prize for his labours in 1922. The principle is that charged molecules (ions) in a vacuum are accelerated by an electric field to a high velocity. Their flight is then deflected by a powerful magnetic field, and the curvature of their trajectory is determined by their charge and their mass. Their acceleration, on the other hand, depends only on their mass, and this is most often determined from the time that they take to reach a detector. In the case of proteins, for instance, special methods were developed to cause ionization. The method is so accurate that if the amino acid sequence, and therefore the molecular weight, is known, it can be used to identify a protein uniquely. Mass spectrometry, it should be said, has many applications beyond the study of macromolecules—in

141

chemistry, medicine (analysis of metabolites, for example, in blood or urine), drug research, and even space research (to discover what molecules exist outside the Earth). Its importance can be gauged from the award of two Nobel Prizes to researchers who refined the principles and improved the instrumentation.

Shapes to order

There are many ways by which chemists manipulate polymer molecules to give them the desired shapes and sizes. Nature provides some clues. Starch, for example, contains, as we have seen, two sugar-based polymers, amylose and amylopectin. The amylopectin is a branched polymer, in which many chains are covalently linked to one another. This is too disordered to form a defined, regular shape, whereas amylose, once separated from the amylopectin, will form partially helical chains. These will associate lengthwise with each other to generate something akin to crystals. It is these interactions that give bread dough and the bread its firm yet yielding texture. Amylopectin, on the other hand, will, given the opportunity, bind much water and turn itself into a gelatinous mass, as will many other disordered chains; gelatin is a familiar case. Chemists and physicists have developed ways of creating to order polymers and polymer mixtures with properties like these, and also much more subtle characteristics. Here now are some examples.

Natural fossil oil is a mixture of hydrocarbon chains of different lengths. It is broken down ('cracked') by distillation at high temperature into a more manageable size range, and distillation then yields a series of fractions: the chains that make up petrol for cars contain around ten carbon atoms, those in aviation fuel are a little longer, while heating oil contains chains averaging about twenty carbons, and lubricating oil about thirty. The viscosity of such hydrocarbon chains increases enormously at low temperatures, and in the Siberian or Alaskan winter engine oil may become almost solid. This always made it difficult to start an engine, and drivers used to resort to desperate measures such as injecting ether into the carburettor with a syringe. But polymer chemists came to the rescue. They devised a random-coil copolymer with many positively and negatively charged groups, yet not so many as to render it insoluble in oil. When added to a light oil of low viscosity it increased the viscosity to the desired level in the cold. In such a charge-bearing polymer the positively and negatively charged groups attract, with the result that it forms a compact coil, for as we have seen (p. 48), polar groups seek to avoid contact with a non-polar medium. Now as the engine approaches its running temperature (at which the viscosity of the pure oil would become too low to afford adequate lubrication), the internal attractions in the polymer diminish. The coils expand, and their contribution to the viscosity rises heftily, offsetting the falling viscosity of the oil itself. With a judicious choice of the proportion of charged groups in the polymer and of its molecular weight and concentration, the solution maintains a practically constant, temperature-independent viscosity.

There are many instances of manipulations of this kind, but let us consider one further example of polymer science, relating to properties of a solid material. Polymeric solids commonly have local crystalline regions, but they may also be totally disordered (amorphous). The polymer chains in either case are packed together as shown below, leaving no free space:

143

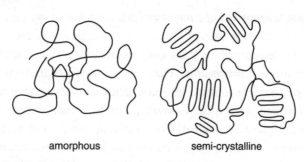

amorphous   semi-crystalline

Figure 30 Amorphous and semi-crystalline polymers.

The resulting material is solid—the chain segments, in other words, can barely move, for if they do they must push away or drag with them the neighbouring molecules. The movement must propagate through the material, an energetically prohibitive proceeding. For most materials, though, there is (in principle) a temperature, not necessarily accessible, at which the higher thermal energy loosens the cohesion of the mass. This is called the *glass transition*, because glass exists in this condition at normal temperatures. Glass is not a liquid at room temperature of course, but it does flow very slowly, which is why ancient windows are thicker at the bottom than the top. In an American laboratory a rod of pitch—a similar material—has been left suspended vertically for many years; about once a year a drop slowly forms at the bottom, and (while an excited crowd gathers) eventually falls. A familiar example of a glass transition is the softening of chewing gum (made most often of gutta percha (p. 110) mixed with a softener, or of polyvinyl acetate) at mouth temperature. The glass transition temperature of cellulose is much higher, which is the reason that ironing of cotton must be done hot.

The plastic kingdoms: thermoplastics and duroplastics

Like the denizens of the plant and animal kingdoms, plastics fall into distinguishable genera. The main subdivisions are thermoplastics and duroplastics. Thermoplastics are those that soften when heated, and so can be formed into the desired shapes, to harden again when cooled. They make up the greater part of the products of the plastics industry, and include predominantly polyethylene, polypropylene, polystyrene, and polyvinyl chloride (PVC). They are not cross-linked (or only very sparsely so), and they are all synthesized by the same type of chemical reaction, for all are made from monomers with carbon–carbon double bonds, which open up to create a chain. (This is not a universal rule, though: some of the less familiar thermoplastics are made by different routes.)

Duroplastics are another matter. They are prepared and moulded at high temperature, so do not soften when reheated. The best known of the duroplastics is Bakelite, and polyurethane is another. Duroplastics owe their characteristic property to extensive cross-linking. This requires chemical ingenuity (even if Leo Baekeland was never quite able to define the mechanism of his own synthetic methods). The simplest strategy is to include among the starting monomers a small proportion of a multifunctional species. Suppose, for instance, that a bifunctional monomer, X—one with two reactive groups, or, as it were, two hands and not just one with which to grip one of its fellows—will generate the polymer $-X-X-X-X-X-X-X-$. If one now includes some trifunctional monomer, Y, one will end up with

Another and more versatile way is to prepare a polymer from monomers that retain a reactive group not involved in the polymerization process, but with affinity for an extraneous bifunctional reagent. When added to the preformed polymer this will react at both ends to effect the cross-linking:

Of course, if the polymer is a floppy coil the reagent can introduce cross-links within a chain. This too can be useful in making a plastic with a chosen character, and conditions can be selected to produce the desired balance between intra-chain and inter-chain bridges.

146

The consequences: strength and weakness, from biscuits to Bakelite

Strength comes from structure, which can be manipulated and fine-tuned in many ways. But when we speak of the strength of a material ('spider silk is stronger than steel'), we need to define the terms with some care. Strength, for a start, is not to be confused with rigidity or stiffness. A biscuit is rigid but weak, while a polyethylene sheet is strong but by no means rigid, whereas iron or steel is both strong and rigid. The most commonly encountered measure of strength is tensile strength—the pulling force required to break a rod of material of unit cross-sectional area. It can be expressed in units such as tons per square inch, but for scientific purposes the unit today is the pascal, written Pa (or more concisely for strong materials, MPa, the megapascal, which is 1 million pascals). On this basis the tensile strength of high-tensile steel is about 2,000 MPa, while that of copper is only 140, wood 100 along the grain but only 3 across; cotton and silk are both about 350, while concrete comes in at a pitiful 4 MPa. At the other end of the scale, nothing compares with the strength of a carbon nanotube (p. 99): that of multi-walled specimens has been estimated as about 100,000 MPa.

Another well-defined property of a material is its susceptibility to distortion—to stretch or bend. These characteristics are defined by a stretching modulus, called Young's modulus (after the English polymath Thomas Young (1773–1829)), and a bending modulus. Young's modulus is the ratio of stress (force per unit cross section) to strain (extension as a fraction of the unstressed length), in the same units as tensile strength. Young's modulus of steel is about 200,000 MPa, while that of glass is 70,000, bone 20,000 and concrete about the same, while rubber is 7 MPa. Again carbon nanotubes take the record: single-walled nanotubes, measuring

147

about 1 nm across (a nanometre is a millionth of a millimetre) have a Young's modulus of around 1 million MPa.

But what counts for many purposes, especially in engineering and building, is *toughness*, or resistance to fracture. This is measured by the speed with which a crack elongates in the material. In a brittle substance that happens with the speed of an explosion. Another critical property is the *impact strength*, or resistance to a blow. The shock wave generated by a blow travels through the material at something like the speed of sound, and can quickly arrive at the weakest point in the structure, which may therefore break at a position quite far from the site of the impact. The worst result will occur in a structural member under tension, so a brittle material, such as concrete, is put under compression (pre-stressed) to minimize the risk of collapse.

Because of the filamentous structure of cellulose (p. 79), wood is very tough, hence its use for mallets, baseball, and cricket bats, and so on. (This attribute is degraded when holes are drilled through the fibres, thus breaking their continuity; the tree maintains the integrity of its wood where a branch erupts through the trunk by filling the hole with a knot, round which the fibres bend without breaking.) Glass, by contrast, is brittle, but fibreglass, which is used for making such stress-resistant items as the hulls of boats, is tough. Fibreglass is a *composite*, produced from two disparate materials, in this case very fine glass fibres, embedded in a plastic, such as a phenolic resin or a polyester. The reason for its toughness is that wherever an advancing crack encounters a boundary with another material its progress is arrested: the sharp tip of the crack spreads outwards and its energy is dissipated. A remarkable example of a composite, ill-starred as it proved, was pykrete, developed during World War II by the eccentric inventor Geoffrey Pyke. The plan was to construct floating airstrips in the

North Atlantic on which aircraft flying between America and Britain with war materials could refuel. Ice is highly brittle—a block of ice struck with a hammer will shatter—but Pyke found that if packed with wood-pulp fibres it was transformed into a medium, pykrete, of enormous toughness. When Pyke demonstrated the properties of his invention before a group of dignitaries by firing a revolver at block of the material, the bullet ricocheted round the room, narrowly missing several spectators. (The plan was abandoned because aircraft with a range sufficient to cross the ocean without refuelling came on stream.)

An advance in the mid-1960s was the development of carbon-fibre materials, which afforded both toughness and great tensile strength. Carbon filaments are made by carbonizing at high temperature threads of a polymer, such as polyacrylonitrile (originally marketed by the Courtauld Company as a clothing fabric under the name of Courtelle). The carbon chains bond to one another in parallel fashion to form microscopic sheets which curl on themselves. These fibres serve as a basis of composites with plastics (or sometimes other materials), which are both light and strong. They find innumerable uses in anything from golf clubs to aircraft. An early failure was in turbine blades of jet engines, which proved to be disastrously vulnerable to bird impact.

Hard biological structures, such as bones, teeth (in which both the hard but relatively brittle enamel of the outer coating and the much softer dentine within consist of inorganic crystals interleaved with layers of protein), and mollusc shells, are all composites and function in the same way. Teeth, in fact, are finely adapted to the diet of the species. Palaeontologists have found that the enamel of teeth belonging to the early hominids is thicker than that in chimpanzees. They infer from this that when our ancestors left the trees some 2 million years ago and took up residence in the savannahs, their eating habits changed from a diet of soft and succulent forest fruits to hard woody roots. The thick enamel layer should help to prevent cracks from starting in the dentine beneath the crown when the tooth crunches down on a tough root or seed, and the enamel layer deforms. The dentine itself is constructed from columns of the mineral matrix and protein, aligned with the axis of the tooth; this makes it more resistant to vertical than to horizontal stresses, just like the grain of wood. Antler bone, finally, is one of the toughest composites—tougher than any man-made ceramic material. The partly crystalline chain alignment of polyethylene and other such thermoplastics likewise favours high resistance to fracture, whereas duroplastics are invariably much more brittle. The toughness of wood is similarly accounted for by the semi-crystalline nature of the aligned cellulose chains.

In the evolution of teeth and other animal structures, though, nature had another obstacle to overcome. Where two materials with different mechanical properties meet—the interface, as it is called—the discontinuity creates the risk of catastrophic failure. For this reason the enamel–dentine interface is a complex layer, no more than $10 \, \mu$m thick, with a hardness that changes in passing from the enamel to the dentine. This enables it to stop the propagation of a crack before it reaches the dentine. The demands are greatest on hard structures which grow out of soft tissue. An example is the beak of the squid, which is one of the hardest known materials, capable of severing the skin, muscle, and backbone of a fish with a snap of its razor-like jaws, and yet is rooted in soft muscle tissue. It is at once obvious that, were its hardness uniform from end to end, the balance of action and reaction would result in almost as much damage to the predator's mouth as to its prey. Nature's solution was to introduce a hardness gradient in the beak, which is soft where it emerges from the muscle and hard as a steel blade at the tip. The creature achieves this by varying the proportions of the chitin (p. 80) and the special protein, which make up the composition of the beak material, and especially the number of chemical cross-links—sparse close to the base and dense towards the tip—which harden the matrix. This type of gradient in composition occurs in many animal appendages with related functions, as for example the fangs of marine worms, which penetrate the tissues of prey and inject a paralysing venom, and also in the threads which anchor a mussel (p. 163) to its rock. These are rooted in the soft muscle of the mollusc, where they are equally soft and compliant, becoming progressively harder and more rigid the closer they are to the rock. The composition gradient is more elaborate than in the squid's beak, for it involves three proteins, all with block-polymer structures,* which assemble into filaments; these also contain ions of metals, such as copper, filtered out of the seawater and selectively assimilated by the proteins. The remarkable adaptive attributes of these materials are, of course, a provocation to materials scientists hoping to simulate their properties in the laboratory. Success would open the way to a new generation of adhesive, coatings, and nano-machines (pp. 205 sqq), with a variety of interesting practical applications.

* A block polymer is one made up of segments differing in composition.

151

8

THE NEW AGE: GIANT MOLECULES FOR THE TWENTY-FIRST CENTURY

A great leap forward in the chemistry and technology of polymers came midway through the twentieth century. It arose out of many failed attempts to prepare a polymer from a monomer that differed from ethylene ($H_2C=CH_2$) only by an additional methyl group:

$$\begin{array}{c} H \\ \diagdown \\ H \diagup \end{array} C = C \begin{array}{c} H \\ \diagup \\ \diagdown CH_3 \end{array}$$

This molecule is propylene. The polypropylenes that resulted when propylene was treated in the same way as ethylene were of very low molecular weight, and mostly sticky gums. The reason for this was that side reactions interrupted the polymerization process. The problem was solved (yet again) by a stroke of the purest luck, combined with an astute response, a perfect example of what Pasteur had in mind when he affirmed that fortune favours only the prepared mind.

Karl Ziegler was a German chemist, whose interests lay in metallo-organic compounds, those in which one or more atoms of a metal are linked to organic (carbon-containing) groups. The syntheses of Ziegler's compounds were carried out in an autoclave, a closed vessel in which the temperature and pressure could be controlled, and in 1953 Ziegler was after compounds made from aluminium and ethylene. To general bafflement, the autoclave was one day found, at the end of the reaction, to contain polyethylene. This was interesting enough to need explaining, especially because the pressure in the autoclave had been quite low, whereas normally a high pressure was required to effect polymerization. But when Ziegler's assistants tried the reaction again they found that it worked on some occasions, but more often not at all. Eventually it transpired that it worked in only one of the laboratory's several autoclaves. Inspection revealed a hairline crack in which were lodged tiny traces of a nickel compound, left from an earlier project. This was the catalyst for low-pressure polymerization. If then this 'Ziegler catalyst', as it became known, was so effective in provoking the transformation of ethylene, might it not also answer for the refractory propylene? So indeed it did, and thus began a whole new era in the plastics industry.

A polymer chemist in Milan, Giulio Natta, who had been working with olefine-based (p. 15) polymers, and especially synthetic rubber, was quick to take note of Ziegler's discovery. Natta had taught himself X-ray crystallography and knew how to determine structures, and when he examined the polypropylene

153

made by Ziegler's method, he discovered something remarkable: the polymer was *stereoregular*. The characteristic of the Ziegler catalyst was that when it set off the polymerization process, the growing chain end would accept only a monomer in the same configuration as that previously, in fact *anti* (p. 136), and so instead of forming a chain with randomly distributed *trans* and positive and negative *gauche* units, an all-*trans* chain resulted. The regular chains have less freedom of movement and can pack together in a crystalline, or at least semi-crystalline manner to make a much stiffer material. Such chains are termed *isotactic*. Natta's work was supported by the huge Milanese chemical concern, Montecatini, and the industrialists were not slow to see the point.

It soon emerged that different types of Ziegler catalyst and different conditions could give rise to two additional forms of the polymer. One was *atactic*, with a conventionally random distribution of configurations along the chain, the other *syndiotactic*, in which the *trans* and *gauche* configurations (the latter either all positive or all negative) alternate. The two regular forms of polypropylene can be represented like this:

while in the atactic form some of the methyl groups are up and some down at random. A better indication of how the chains look in space is given by the representation below, in which R, as before, can be a methyl group, as in polypropylene, or any other substituent. Here the wedge-shaped bonds project out of the page, and the dashed bonds point back from the page:

Because the regularity of the atactic and syndiotactic chains imposes a fixed angle of twist between successive links, they assume a helical rather than a randomly coiled structure. But because none of the units are intrinsically asymmetric, in the sense that for instance an L- or a D-amino acid (p. 20) is, there is nothing to dictate whether the helix should have a left- or right-handed screw sense, and therefore left- and right-handed helices occur indiscriminately in the solid. Atactic polypropylene is of little practical use, but its stereoregular cousins have a vast range of applications. Polypropylene is a thermoplastic, moulded into rigid or flexible containers for many purposes, including laboratory ware, for it is resistant to solvents and most other chemicals, as well as car, aircraft, and electronic parts. It makes excellent packaging for food, and finds uses in textiles and even in banknotes. A rubbery form of polypropylene is also manufactured by suitable manipulation of the polymerization conditions.

Acrimony developed between Ziegler and Natta over questions of priority, but must have been assuaged by the award of a joint Nobel Prize in 1963. It is worth mentioning that, as in so much of the history of large molecules, the discovery of isotacticity was anticipated by Nature. In 1925 Maurice Lemoigne, a French biologist, isolated from bacteria a polyester—the first ever seen— polyhydroxybutyrate (PHB); the bacteria apparently synthesize the substance under conditions of environmental stress. Later several other related natural polymers were discovered. The PHB can be turned into a heat-resistant, largely isotactic thermoplast with many desirable properties. The most attractive of these came to light when commercial laboratories began to take an interest, during the dawn of the environmental movement, for PHB, unlike the run of plastics, was biodegradable. A university laboratory in the United States incorporated into a plant and into the favourite laboratory bacterium, *Escherichia coli*, the genes which the bacteria use to synthesize PHB. The polymer could now be produced on a sizeable scale. The Monsanto Company, which had acquired the patent rights, marketed PHB under the trade name of Biopol, but with the advent of genetically modified crops, Monsanto evidently decided it had bigger fish to fry, and sold out to a smaller outfit. It remains to be seen whether anything practical ever comes of PHB or its analogues.

Virtues of rigidity

Double bonds, as we have seen, severely restrict the freedom of motion of the polymer chain. In actuality the nature of covalent chemical bonds is more nuanced than the simple categories of single and double bonds implies. The electrons which form a bond between atoms of grossly different character are in general asymmetrically disposed, with a bias towards the more electropositive atom. And in a chain of identical atoms (carbon), but with alternating single and double bonds between them, or in a ring in which this occurs, like benzene (p. 16), the electrons are 'delocalized'. There is, in other words, no difference between successive bonds, even though they are written as alternating single and double bonds in the formula. Thus all six bonds in the benzene ring have the same length (whereas a true double bond is shorter, by reason of the tighter association between the atoms, than a single bond); they are all identical and midway between a single and a double bond. The electrons in fact form a homogeneous cloud around the ring. This phenomenon—also seen in linear chains with alternating single and double bonds—is referred to as *conjugation*. Now among the bonds written in the formulae as single bonds that in reality have partial double-bond character is the amide bond,

$$\underset{\diagup}{\overset{\diagdown}{C}} - \underset{}{\overset{\overset{\textstyle O}{\|}}{C}} \cdots \underset{\diagdown}{\overset{\diagup}{N}} - H$$

as indicated by the dashed line. This forms the backbone of protein chains (p. 31) and of the synthetic polyamides, such as nylon, and greatly restricts the structural freedom of the chain. Nylon normally exists in a semi-crystalline state, with islands of chains all stuck together by hydrogen bonding, very like the β-structure of proteins, and lying roughly in the direction of the fibre axis, in a sea of floppy, disorganized chains.

157

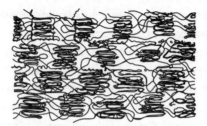

Figure 31 A typical polyamide, such as nylon, in which the chain makes a pattern of semi-crystalline islands, interspersed with floppy unstructured segments.

Spinning or drawing threads of the polymer causes partial elongation of the chains, with an increase in tensile strength.

A particularly rigid form of the polyamide chain was developed in the 1960s. It is a nylon with benzene groups, which are bulky enough to impede twisting motions of the chain, and so, unlike the polypeptide chain of proteins, cannot form a helix. Moreover, it is in the nature of amide groups to associate with each other through hydrogen bonds (as we have seen in the α-helix (p. 35)). Such bonds, individually weak, can generate cumulatively strong, interactions. The formula of this polymer is:

and it is called polyphenyleneterephthalamide, or more familiarly *kevlar*. It is chemically highly resistant, and soluble only in strong acids or polymer melts. When dissolved in such a medium the molecules are more or less rod-like, but when spun from solutions the rods align in the direction of pulling, like tree trunks in a flowing river, and form extremely strong fibres. The polymer is relatively elastic, and of course light, since it consists only of light-weight atoms, but extremely tough—weight-for-weight about five times more so than steel. Kevlar has found many uses—in ropes that need to bear a high tension, in tyre cords, in the hulls of small boats, and most famously in body armour and helmets. Bullet-proof vests are made of cross-woven kevlar fibres, which absorb much of the energy when an impact forces them to bend, and dissipate it further by distributing it over a wide area around the region of the blow.

Kevlar is only the first of a number of polymers of like character, and some of the others, stiff, rigid yet elastic, are used to make light and effective armour plating for military vehicles. Carbon nanotubes have been introduced into polymers to create composites of exceptional toughness and resilience. Body armour made from nanotubes may follow (although the cost is now still too high). It should resist penetration by a bullet, but the force of the high-velocity impact would probably cause fractures and haemorrhages. It has even been mooted that nanotubes could be used in the detection of cracks in structures, such as turbine blades and aircraft wings, because the heat conductivity of the structure, dependent on the nanotubes, would be reduced in the vicinity of the crack. Not only that, but if a heat-activated adhesive were incorporated in the material, the heat conductivity of the nanotubes would cause local heating and mend the crack.

Branches and dendrimers

Polymer structures are not confined to linear chains. Some biological polymers, of which glycogen, the body's sugar store, is one, are highly branched (p. 81). A linear polymer with reactive side chains, such as a protein, can be made to grow whisker-like appendages. Proteins contain the amino acid lysine, which displays an amino group, $-NH_2$, at the end furthest from the backbone. Under normal conditions the amino group carries an extra proton (p. 31), so has the structure $-NH_3^+$. Such charged groups are located at the outer surface of globular proteins, in contact with water, and present a good target for chemical reactions. The amino group is reactive, and can serve as an initiator of a polymerization reaction. Depending on the number of lysines in the chain, the surface of the protein can be more or less densely coated with whiskers of, for example, polyethylene glycol,

$$\diagup CH_2 \diagdown_{CH_2} \diagup^O \diagdown_{CH_2} \diagup^{CH_2} \diagdown \diagup^{CH_2} \diagdown_{CH_2} \diagup^O \diagdown_{CH_2} \diagup^{CH_2} \diagdown_O \diagup$$

Polyethylene glycol, commonly referred to as PEG, is made from the monomer, ethylene glycol,

$$H_2C-OH$$
$$|$$
$$H_2C-OH$$

a common chemical familiar as antifreeze for car engines, and in certain infamous cases as an adulterant in wine, to add sweetness and unction. PEG is one of a large group of polymers, called *polyethers* (an ether being a compound of the type $R-O-R'$, in which R and R' are alkyl groups as before; the anaesthetic ether is

160

properly called diethyl ether, for both R and R′ are ethyl groups, C_2H_5-. PEG, has many uses by itself—in the food industry as a dispersant to preserve emulsions and foams, in medicine (as a laxative), and in cosmetics and toothpastes.

Attachment of PEG to proteins (pegylation for short) affords a way of making refractory proteins water-soluble, or masking their antigenic properties, but such uses have so far been very limited. Much more interesting is the class of polymers termed *dendrimers* (from the Greek for tree, because they have a branched structure). Dendrimers can be built up in stages from a multi-valent core, and have a structure like this (though with many variations):

Figure 32 Schematic picture of a dendrimer. The lines are polymer molecules with branches. The number of branches increases going outwards from the core. The circles indicate the successive steps in branch formation.

The separate branches emanating from the core are called dendrons, and dendrimers are also sometimes synthesized by linking preformed dendrons to one another. We will come to the useful applications of dendrimers a little later.

9

PRESENT AND FUTURE: WHERE WILL IT END?

We have arrived now at the almost limitless ways in which polymers, natural and synthetic, enter, often unseen, into the fabric of modern life, and of what they may do for us in the future. Many of the most enterprising developments come from the scrutiny of the materials and processes of life, for why should we not look for inspiration from the products of aeons of evolutionary trial and error? Biology-inspired chemistry and physics has been given the name biomimetics, and we will encounter many examples of its successes, and indications of its widening promise.

Adhesives and tacks

Giant molecules have served as adhesives since ancient times. In earlier millennia they were of course derived from natural substances. First probably (in Egypt from about the fourth millennium BC) came vegetable starches, boiled up in water. Then there were glues made of collagen, the protein extracted as gelatin from bones and the skin of mammals and fishes (p. 43), and much later the cellulose derivatives and rubber solutions. All but the last have been supplanted now by synthetic polymers, in general duroplastics, most often elastomers above their glass transition (p. 144). Among these are the epoxy or cyanoacrylate resins (Araldite, Superglue, etc). They are 'hardened' by cross-linking, and the trick is to get the extent of the reaction just right, for if there are too few cross-links the adhesive remains viscous and will flow when stressed, if too many, it becomes rigid and brittle. In between it will respond elastically to stress, and cracks will not propagate.

Remarkable examples of fine-tuning of cross-linking reactions occur in nature, especially in marine molluscs, such as mussels, barnacles, and limpets. Their glues (called 'holdfasts') work under water—something that science has only recently begun to emulate. The blue mussel attaches itself to a rock through a small appendage, or foot. Two tiny orifices exude a protein gel and a hardening solution. Some five to ten proteins are involved, but the principal cross-linking agent is an amino acid known as L-DOPA, familiar as a neurotransmitter in mammals and used in the treatment of such conditions as Parkinson's disease. The adhesive proteins are small as proteins go (little more than 100 amino acids), and have an unusual composition, with many charged amino acids. The cross-links are very precisely disposed, and at least one of the components serves to impose a fixed separation on the protein filaments. The whole process occurs with great despatch: it takes only minutes for the mussel to expel and mix the glue, and hardening is complete in less than a day. The preponderant protein has been produced in the laboratory by genetic engineering in bacteria, and current work aims at repro-ducing by arduous technology what the mollusc achieves with no discernible effort.

The primary requirement for adhesion, generally, is a high propensity of the surface to be wetted by the adhesive fluid, which must spread, like water on clean glass, and not run off like drops of water on Teflon. A polymer surface that changes its state from crystalline to fluid and viscous in response to a stimulus, such as a small rise in temperature, can permit attachment or detachment on demand. Suppose, for instance, that the polymer is made from two monomers of different character, one with a low, the other with a high surface energy. (This is equivalent to surface tension, the inward pull of the molecules at a surface, which causes a liquid to cohere into a drop.) Then during crystallization, which occurs at low temperatures, the groups of low surface energy will tend to gather at the surface in contact only with air. This means that the surface after crystallization will have little inclination to wet and adhere. Above the crystallization temperature the polymer segments are free to move about, and so the groups with the high surface energy will tend to concentrate at the interface, there to encounter the adhesive. Other ways of controlling an adhesive interaction, as by a change in acidity or pressure, have also been developed, and are of interest in the cosmetics industry and else-where. Post-it labels and self-sealing envelopes are examples of reversible pressure-controlled tacks.

Adhesion can sometimes be promoted by preparing the surface, depending on its chemical nature, before application of the glue. If the surface will support polymerization—in other words initiate the reaction of monomers to form a polymer—the resulting poly-mer molecules can become entangled with the polymer chains of the glue, thereby greatly strengthening the attachment. This simulates on the molecular scale the entanglement responsible for the reversible seal between Velcro surfaces. Velcro (which, as many know, was invented by a Swiss engineer ruminating on the retention of burrs by his trousers after a hike) is most often made from nylon. Nylon fabrics can produce, when worked in a machine, a surface roughened by minute filaments ending in hooks. When two such surfaces are brought into close contact the hooks engage and bond the surfaces together.

Now, obviously, the more attachment sites there are on a surface the better it will stick, not only because the strengths of the individual interactions add up, but also because the surrounding contacts ensure that if any one, or even a whole patch should break, the gap can quickly reseal. The gecko lizard can climb up a glass wall, or even walk upside down on a glass surface. This is because it has millions of adhesive hairs on its footpads, each of which binds only weakly, but together they exert a powerful force of attachment. The force exerted per unit area of surface by a single hair has been measured, and in the standard units (actually N/cm^2, where N stands for the Newton, the unit of force—though we are not concerned here with absolutes) it is one ten-millionth; but the foot pad with its myriad hairs exerts a force one-hundred-thousand million times greater (10^7, as against $10^{-7} N/cm^2$). The diameter of the hair, it should be added, is also important in determining the adhesive force. Polymer sheets, or tapes, with hair-like protrusions like the gecko's have been made by microfabrication techniques, and have the same powerful adhesive properties, in effect a super-Velcro. Another group of researchers have taken this concept a step further, for the polymer tapes, like the gecko itself, do not manage nearly so well in water. Their microfabricated polymer surface was furnished with a thin coat of a different, water-insoluble, polymer, modelled to

165

resemble the 'holdfast' protein of the mussel (rich in L-DOPA). The result is a surface which retains its adhesive properties almost undiminished under water.

Abhesives and other useful coatings

A substance that prevents adhesion is sometimes referred to by surface scientists as an abhesive. Polymers with this property (like Teflon) have great industrial importance, notably as antifouling agents on the hulls of ships. Their use has been proposed for windscreens and for windows of houses to prevent accretion of dead insects, dust particles and other detritus. Rust-proofing is an equally profitable undertaking for which a polymer has been developed. Rust is an iron oxide, and results from the reaction of the iron with atmospheric oxygen. A traditional way to shield iron against rusting is to coat it with metallic zinc. Zinc is more electropositive than iron, and so (recalling (p. 49) that oxidation amounts to loss of electrons) it, rather than iron, becomes the preferred source of the electrons taken up by the reactive oxygen. But zinc is expensive and not wholly effective or durable, and therefore a polymer, which serves the same function much more cheaply, is coming into use. This is polyaniline (aniline being nothing more than aminobenzene—a benzene ring bearing a single $-NH_2$ group). It acts as an 'organic metal', an intermediary which accepts electrons from the iron atom and transfers them to oxygen. A thin layer of pure, inert iron oxide forms beneath the coating and arrests the corrosion process, perhaps for ever.

Another objective hotly pursued by polymer chemists is a polymeric coating for buildings which will function like a solar panel and absorb light energy. This requires a polymer that conducts electricity—and some such already exist, as we shall see—but does not radiate the light back as fluorescence. Problems remain to be solved, such as averting the eventual photochemical breakdown of the polymer, but prototypes have been developed. A less ambitious application might be in solar-powered computers. Polymers which do not reflect radar waves have also been developed, and are perhaps already in use on the battlefield as coatings for vehicles.

An altogether different type of coating comes from mushrooms and other fungi. The *hydrophobins* were discovered in about 1990. They are proteins which help the fungus to thrive under a variety of often unpropitious conditions, and they act apparently by creating a water-repellent layer on vulnerable parts of the organism, while at the same time promoting adhesion to symbiotic plants. There are two classes of hydrophobins, one totally insoluble, the other with selective solubility under certain conditions. Both possess the remarkable property of self-assembly, of spontaneously coming together, in other words, to form *amphipathic* membranes. The term implies that one side of the membrane sheet, one molecule thick, is polar (p. 60), therefore hydrophilic, the other nonpolar and hydrophobic. One side will consequently tend to adhere to most surfaces, thereby exposing the other, non-adhesive side to the environment. The hydrophobins are small proteins and can be produced in quantity, modified if required, in genetically engineered bacteria or plants. A wide range of uses is envisaged and some have been tested. The hydrophobins may one day supplant existing antifouling agents on sea and land.

It has already been shown that the hydrophobin membrane surface supports the growth and fusion of skin cells, and so may prove a superior material for the manufacture of surgical

167

implants. The membranes may also be used to lend stability to biosensors—devices that use immobilized enzymes or other biological molecules to detect and quantify metabolites and other substances. The hydrophobins may find applications in the ever-voracious and inventive food industry: they may serve, like agar and carrageenans (p. 82), to stabilize emulsions, such as ice creams and mousses. They may further be expected to render layers of creamy unguents and oils more resistant to water and detergents, which brings us directly to the cosmetics industry, an alluring playground for polymer chemists.

Beauty and the polymer

Hair styling requires fixatives. Pomades, made of coloured and scented hydrocarbons, have largely gone the way of moustache wax, nor is gum tragacanth, a water-soluble glue extracted from the sap of a species of tree, still to be found in the barber's shop. The demand now is for something less tacky, most often a water-soluble polymer, polyvinylpyrrolidone, or a related copolymer. This has a limited lifetime on the hair, especially under humid conditions, and so gelling agents, as well as resins, delivered in aerosol sprays, fill the supermarket shelves. Any number of polymers, copolymers, and block polymers and their mixtures are in use as coating, gelling, and fixing agents. In general, a glass transition at relatively high temperature is desirable, and another valued property is good surface reflectance to impart a high gloss.

These marvels are all far removed, of course, from operations on the hair, favoured during most of the past century, chief of which was 'the perm', or permanent waving. To mould the hair in this way requires chemical modification of its substance, the protein keratin (p. 37). The first recorded attempts date back to about 1900, when a German hairdresser, Karl Neissler, experimented with some fortitude on his wife. It is said that his initial efforts involved treatment with bovine urine, but it is not clear what this could have achieved. He went on, though, to twist the hair round rollers and apply so much heat that the coiffure was burned down to stubble. Success of a kind followed when Neissler treated the hair with strong alkali (sodium hydroxide) at round about boiling-water temperature. The consequences for the condition of the hair must not have been good. Nevertheless, Neissler developed a heating apparatus which found its way into the best salons in Europe and America.

The keratin chains are cross-linked by disulphide bridges (p. 49), which give the hair its elasticity, and in normal circumstances preserve its form. The disulphide bonds, $-S-S-$, are easily ruptured by a reducing agent (p. 49), giving place to two sulphydryl groups, $-SH$. If now a new shape is imposed on the hair with the aid of clamps or rollers and softening by heat, oxidation will cause disulphide bridges to reform, but between different pairs of sulphydryl groups, newly brought into apposition by the deformation. The new configuration is now stable. The reducing agent first used was ammonium thioglycolate, and the oxidizing agent hydrogen peroxide (at a concentration low enough, no doubt, to avoid turning all the clients into platinum blondes). This technique came into use in 1938, the treatment lasting 6–8 hours. Better reducing agents, and modified conditions, especially control of the pH, brought the time of treatment down to about 30 minutes or even less. In the interim there were many audacious experiments with chemicals and high temperatures, at least one of which caused the hair to fuse into a kind of horny helmet, which could be lifted from the denuded scalp beneath. Many lawsuits ensued. The cleavage and reformation of disulphide bonds could equally well allow curly or crinkly hair to be straightened, a fashionable transformation in some parts of Africa.

Returning now to our own time, conditioners in shampoos are a busy field for polymer chemists. The keratin of the hair carries a negative charge (like the majority of proteins), and so a positively charged polymer will tend to stick to it. All manner of polymers, simple and complex (some charged, some neutral and based on silicon–oxygen chains (p. 3), others possessing both these and carbon-based chains), are common ingredients. Polymer science is similarly inseparable from formulations of lipsticks (although beeswax remains a widely favoured base), mascara, and face creams. These are generally thickened with polyethylene glycol, carrageenan from seaweed, or more complex synthetic macromolecules. But enough; we must move now from vanity and luxury to necessities for life.

Desalination

Water shortages have been the cause of poverty, disease, starvation, and wars. Endless time, ingenuity, and treasure have gone into finding ways to purify polluted waters and eliminate the salt from seawater. Polymer membranes with pores of a size to allow passage of water but not of bacteria, or viruses, are now commonplace. A domestic handheld device has even been invented in which one pumps water through a tubular membrane to filter out contaminants. The great problem has always been fouling of the membranes, mainly by organic matter, especially sticky polysaccharides exuded by bacteria and other micro-organisms. This set off a search for fouling-resistant membranes, made for instance by grafting hydrophilic whiskers onto the hydrophobic, porous matrix.

Eliminating salt, the constituents of which (sodium and chloride ions) are smaller than water molecules, is much harder, and the problem is becoming ever more urgent as fresh-water supplies dwindle. (At best they make up no more than about 2.5% of the planet's water.) Separating the parts of a mixture means working against entropy (p. 115), which cannot be done without expenditure of energy, and the aim of desalination, and of water-purification research generally, has been to increase energy efficiency. Salty water can be passed through a column—a long tube—containing a polymer resin, called an ion-exchanger. The polymer carries a high density of electrostatically charged groups, to which salts and most other impurities will bind. This works well enough, but the material soon becomes saturated with the ions of the salt that it has removed; then it must then be regenerated, and to do this water is needed, which gets one back to where one started. Nearly all desalination plants operate on the principle of *reverse osmosis*, in which salt water is driven at high pressure through a polymer membrane that allows water to pass, while rejecting salts. Energy, of course, is consumed by the pumps that maintain the pressure, and the efficiency of the process is limited: on the one hand, thick, viscous, highly concentrated brine accumulates behind the membrane, and on the other, one is again up against fouling by suspended matter. Attempts to improve efficiency largely centre on microfabrication of membranes with pores of uniform size, and attempts are even in hand to embed carbon nanotubes of appropriate diameter to admit water in a membrane matrix.

171

There is a school of thought which holds that the future of desalination lies in biomimetics—perhaps in the incorporation into a polymer membrane of water-selective protein channels. Such channels are found in the membranes of most animal cells: most consist of a protein called aquaporin. The first step towards the construction of a working membrane based on this approach, through which water will pass without application of high pressure, is to integrate the protein in active form in the polymer matrix, and that has already been achieved. This scheme could also be stood on its head: if an ion-pump protein can be incorporated in a membrane, then with ATP as fuel (p. 62), it will allow the membrane to retain the water while pumping out the ions. Developing such processes on an industrial scale will be no trivial matter, but the idea is by no means fanciful, for proteins, if they are not too large, can now be prepared on a huge scale in genetically engineered bacteria or plants, or in the milk of genetically modified domestic animals.

The polymers used for this and similar complexes with protein channels or pumps are amphiphilic block copolymers. They are constructed with alternating blocks of hydrophobic plastic-like segments, such as polystyrene, and charged, or at least hydrophilic, segments. A properly designed polymer of this nature forms itself into a vesicle, a membrane-enclosed bags of water, just as lipids do (p. 61), but being covalently bonded, the polymer vesicles have a far greater stability. The proteins, such as aquaporin, usually reside in the cell membrane (made of lipid), and therefore sit comfortably within the hydrophilic regions (though this makes the creation of such vesicles appear too easy). It will be apparent that desalination is only one of the possible applications of such hybrid membranes: controlled drug delivery is another.

Delivery vehicles

To function properly most drugs must be provided to the patient in a controlled amount and over a predetermined, often extended, period. One way to achieve this is with a pump delivering at a set rate into a vein. The disadvantages are obvious, and the danger of infection perhaps the greatest. For this reason liposomes have aroused high hopes. Liposomes are small bubbles of water enclosed by a membrane. The membranes of the cells in our body are made predominantly of phospholipids, together with cholesterol. A phospholipid is a fat, made up of a phosphate group, to which are attached two long hydrocarbon tails (p. 61). Hydrocarbons are, as we have seen, quintessentially hydrophobic, and avoid exposure to water. When shaken with water they form an emulsion, with spherical droplets, for the sphere has the small-est possible ratio of surface to volume. The hydrocarbon chains attached to the charged, therefore highly hydrophilic, phosphate group can do better. They cluster together in a so-called bilayer— a sheet of two layers with hydrocarbon chains in contact on the inner surfaces, while both outer surfaces bristle with the phos-phate groups. The hydrocarbons are thereby shielded from the water. But at the edges of a flat bilayer sheet the hydrocarbons would still be exposed to the watery environment, and there-fore the sheets curl up to form a closed bag containing water. Natural cell membranes (p. 61) also contain embedded proteins which are channels of communication between the interior of the cell and the external medium, but to make a rudimentary artifi-cial cell—a liposome, or vesicle—only phospholipids are needed. This is how they look, with phosphate groups depicted as black spheres:

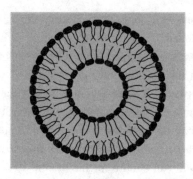

Figure 33 A liposome, or phospolipid vesicle. The black blobs are the charged phosphate groups and the lines are the hydrocarbon chains.

Liposomes are easily made, are tolerated by the body, for they are composed of normal body constituents, and can encapsulate drugs. They have been used with some success, but they do not last long in the circulation, and they tend also to attach themselves unprofitably to cells.

Also of interest for the delivery of drugs, and especially for the introduction of foreign genes into cells for experimental therapies, are viral capsids. These are in essence viruses without their nucleic acid. Most viruses consist of nothing more than a shell of protein molecules, sheltering the nucleic acid (DNA or RNA) within. When the virus attacks a cell, a fusion event occurs between the virus shell and the cell's outer membrane. This allows the nucleic acid to pass into the cell and take command of its metabolic machinery. A piece of DNA (in general a gene) contained in a virus shell emptied of its own nucleic acid can therefore enter a target cell in the same manner. The viral proteins assemble spontaneously into the shell structure, so nucleic acids or drugs can be easily encapsulated.

Polymer cages with long survival times in the body have also been developed. They may take the form of a cross-linked polymer gel with drugs trapped in its interstices. (In the absence of cross-links, a hydrophilic polymer of this kind would simply dissolve in the blood.) The drug molecules diffuse out into the blood plasma at a rate determined by the porosity of the gel. The capsules may be swallowed, injected, or implanted under the skin. This last technique has been used to deliver a contraceptive agent, and can endure for months or even years, slowly emitting the tiny amounts of drug required. The material is most commonly a copolymer constructed from two natural metabolites, lactic and glycolic acids, which are eventually degraded by the body's enzymes. To direct these 'microspheres' (or nanoparticles or nanospheres) to a predetermined location, such as a tumour, the polymer is chemically coupled to an antibody, which recognizes and fastens itself onto a tumour-associated protein (antigen). Many such 'tumour markers'—proteins which reveal themselves on the malignant, but not on the surrounding healthy cells—are known. Instead of an antibody, another type of 'magic bullet' may be used to direct the drug-laden nanosphere to the site of action. This is an *aptamer*—a short piece of nucleic acid, DNA or RNA, with a sequence that recognizes the tumour-specific molecule.

More refined polymer cages are the so-called *polymersomes*. These form spontaneously from water-soluble copolymers consisting of segments of hydrophobic and hydrophilic (usually charged) constituents. The nature and lengths of the two types of segment and the overall length of the copolymer determine the size, shape, and thickness of the self-assembled shell. Because of their similarity to natural cell membranes, these little bags can merge into the surface of, for instance, a tumour cell, and there

175

release their contents, in general an anti-tumour drug. With chemical expertise, the lifetime of a polymersome in the circulation can also be tuned, if its contents are to be dispersed at large in the body.

Activity in the drug delivery business also now centres on another type of system, the dendrimer (p. 161), within which foreign molecules can be distributed like currants in a bun. Dendrimers possess several advantages. Drug molecules trapped in or close to the central core may emerge very slowly or not at all, while those closer to the surface, where the crevices are larger, will diffuse out more rapidly. The dendrimer can be tailored to accommodate a drug of a given size and show the desired pattern of release—starting rapidly and tailing off. But some drugs also work while remaining attached to the giant molecular matrix, and the dendrimer affords a huge number of attachment sites, supposing that the drug molecule can be chemically coupled to the tips (terminal groups) of the many branches. A dendrimer thus studded with drug molecules would be especially advantageous when the drugs are insoluble in bodily fluids. Where the high capacity has already proved its value is in imaging by CAT scanners and the like, because the very high concentration of the radiation emitters, gathered together in the dendrimer, makes for greatly enhanced sensitivity of detection by the scanner. The compact nature of the dendrimer has the further virtue in this type of application, denied to linear and therefore much more expanded polymer molecules, that it is cleared in relatively quick time by the kidneys.

In the longer term, carbon nanotubes could become the delivery vehicles of choice, with the drug inside the tube (the domed end removed) so as to diffuse out at a rate determined by the diameter of the tube; or it could be attached by a labile link to the surface. Chemical modification of the carbon to allow attachment is relatively simple (depending on the chemical nature of the drug), and if multiwalled nanotubes are used the capacity would be very large. The scope of nanotubes for therapy can be gleaned from the following spectacular demonstration: cancer cells differ, as indicated earlier, from the normal tissue surrounding them in the proteins that they display on their surface, commonly receptors (p. 62) for metabolic molecules. Certain tumours have a high density of receptors for the vitamin, folic acid. Researchers in the United States bound folic acid to nanotubes, which accordingly attached themselves to tumour cells and to no others. The plan was to take advantage of a useful property of the graphene—its propensity to absorb light in the near-infrared region of the spectrum (to the long-wavelength side of the visible range). Absorption of infrared radiation causes heating, so irradiating the system with an infrared laser—a rudimentary device, no bigger than a pocket torch—raised the temperature of the nanotubes, but not of the surrounding tissue. The result was that the cells to which they were grappled through the folic acid were warmed as if by tiny heating elements, and within a minute or two reached the lethal temperature of 70 °C. Thus the tumour cells died, while the normal cells were untouched. This experiment, it has to be said, was performed only on cells in water, and it remains to be seen how the method can be made to work on solid tumours in the body.

Tissue and organ engineering: the plastic pancreas

Tissue engineering is today an established part of clinical research and indeed practice. It began haltingly with the use of animal skin to cover parts of the body area of patients with severe burns.

177

This was a desperate measure because of the twin dangers of immune catastrophe and infection. Now a few of the patient's own skin cells (to avert rejection) can be 'seeded' onto a matrix on which they will grow and spread. When nicely covered, the sheet is applied to the burn or other wound. A biocompatible matrix, referred to in the bioengineering trade as a *scaffold*, is indispensable. Biocompatibility means that the material will not attract blood cells and proteins, which invariably home in on any unfriendly surface and form massive accretions—something that can happen in minutes. Nor must the material be identified as alien and induce an immune reaction. Natural proteins, or at least those seen as 'self' rather than alien, are biocompatible, and so are certain hydrophilic copolymers, such as that already mentioned, consisting of lactic acid and glycolic acid units (PLGA for short). This and other biocompatible polymers are produced from monomers which occur in the body, or are sufficiently similar to ones that do. It is also desirable that implanted scaffolds should spontaneously degrade during the time the cells need to form a continuous sheet over the wound area and become integrated into a patient's existing tissue. The PLGA is indeed slowly broken down by degradative enzymes into carbon dioxide and water. This is why it also finds a use in surgical sutures, along with more elaborately engineered materials. One of those is avowedly based on the principle of the gecko's foot. A tough biodegradable copolymer sheet was engineered to present a pattern of tiny projections on the scale of the gecko's hairs (p. 165). It was formed by photolithography (see later), and the resulting prickly sheet was coated with a polymer of sugar units chemically modified to form chemical bonds with the tissue beneath. The begetters of this creation envisage adding more components to the surface, such as growth factors or other substances known to promote healing.

But to grow, divide, and cover the area of the wound the cells also prefer a surface of the right texture, resembling in fact that of the *extracellular matrix*, the natural substrate on which skin cells (and many others) are programmed to grow. This means that the mesh of protein filaments coating the extracellular matrix must be simulated in the laboratory. If the fibres are too narrow the cells will not adhere properly; if too wide, they will flatten, spread out, and remain inert. There are several ways in which this problem has been tackled. One is to spin the biocompatible polymer into fibres of the right thickness, which are then applied to a sheet of supporting material. Another option is to generate the precisely patterned surface structure by CAD/CAM—computer-assisted design and manufacture—using a programmed device like an inkjet printer to deposit rows of tiny blobs of polymer. An aid to cell growth is to incorporate in the matrix polymer the simple and universal motif of three amino acids (arginine–glycine–aspartic acid) recognized by a group of special adhesion proteins, called integrins, which reside in the cell membrane.

Cell growth, of course, is a slow process, but the long wait can often be reduced by making the patient's own cells grow inwards from the edges of the wound. In this procedure an 'artificial skin' is made from two layers—an epidermis (in natural skin, the protective outer protein layer) comprising an elastic water-resistant silicon-containing polymer, and a dermis made up of fibres of natural skin constituents. These are the principal skin protein, collagen (p. 43), and the glycosaminoglycan chondroitin sulphate (p. 82). When this patch is placed over the wound the skin cells at the edges advance into it, for they have the innate propensity to crawl along a compatible surface and fill up all available space. When the dermis has

179

fully mended, the outer polymer film is surgically peeled away, and replaced by a very thin layer of the patient's transplanted epidermis.

There has been much research on possible ways of extending this kind of technique to whole organs, laying down cells by CAD/CAM in a three-dimensional array. Meanwhile simpler implants to act as an artificial pancreas or liver have been developed. The surrogate pancreas, for instance, has the form of a tube made of several polymers. Pancreatic islet cells—those which produce insulin when stimulated by glucose—from a human donor or an animal are embedded in a soft, porous polymer matrix. The matrix is permeated by fibres of a stiffer polymer to lend it mechanical strength. This sausage is encased in a selectively permeable membrane, which allows access to sugars and other small metabolites, but not to cells or to antibodies, so the pancreatic cells remain hidden from the body's immune system. And finally, over the whole there is a porous biocompatible outer sheath of PLGA or a like material. Artificial livers have been made in the same way. These organs can survive and function in the body for months or years. There is also much interest in the creation of artificial blood vessels, especially arteries, to replace those that have ceased to function, or threaten to deteriorate. Prototypes have been fabricated from strong, sufficiently elastic fibres of a stable plastic, such as Dacron, and the interstices sealed with a natural protein. When it comes to circulation the problems of achieving biocompatibility are at their most challenging, and a strategy that has been tried (with what success is not yet clear) is chemically attaching heparin to the inner surface of the vessel. Heparin is a natural sugar-type polymer which sequesters calcium ions, and thus prevents the formation of blood clots, as would otherwise occur.

Artificial skin for permanent wear is a different matter. Here the idea is to provide a covering for prosthetic limbs, which would feel and function like natural skin. A possible approach to this ambitious goal starts from a durable, flexible polymer (of which many are available) containing carbon nanotubes. This material will conduct heat so could restore the sensation of hot and cold. Because nerves can now be led from other parts of the body, it is proposed to incorporate arrays of pressure sensors into the skin, and thus provide a sense of touch.

Replacement or repair of bones presents its own peculiar problems. Under mechanical stress bone cells, called osteoblasts, make new bone (whereas underused bone slowly atrophies—a problem of concern to astronauts and the bedridden). Metal (generally titanium) implants are stronger than bone and thus absorb applied stresses, so that the osteoblasts remain dormant. The aim therefore is to create polymers for implants that will have tensile strength similar to that of bone, and will also of course be biocompatible. To assist the healing of fractures, the broken surfaces need only to be locked in contact. For bones not required to bear high tensile loads, this can be achieved by biocompatible and biodegradable polymers, and here again PGLA fits the bill, and is widely used. No doubt the next few years will see new polymers, designed to meet specific needs in the body, come on stream.

Smart polymers

'Smart' (or intelligent) materials are in general polymers, designed from scratch or modified from natural molecules, with the capacity to respond to a change in their environment. The signal to change shape or adjust some other property may be a rise or fall in temperature, light intensity, or pH (that is acidity), or the

181

arrival of an electrical impulse. A familiar smart material is the photochromic spectacle lens, which darkens when the light gets brighter. The natural world is replete with smart materials—the bacterial flagellum, the whip-like appendage which rows the bacterium towards the source of nutrient, our muscles which contract on release of calcium ions from the nerve terminal, the lens that changes its focal length when your gaze shifts from the page to the horizon, the retina that records an image of what passes through that lens and despatches it in encoded form to the brain, and so on. In all cases it is a combination of macromolecules—proteins—that do the trick.

The structural elements—mollusc shells, bones, skin, connective tissue, and the rest—are all, as we have seen, composites (which materials scientists are trying to simulate in the laboratory (p. 45)). One of the oddest is the skin of such creatures as the sea cucumber. This is in reality a type of connective tissue, rather than a true skin, and it has the remarkable property of changing instantaneously when the creature is disturbed, from an almost fluid to a stiff texture. The sea cucumber can squeeze through a crevice in a rock when its skin is fluid-like and viscous, but will resist efforts by a predator to pull it out by turning rigid. The material is called catch, or mutable, connective tissue, and it operates through a reversible cross-linking mechanism. It has a layered structure, the bulk of which consists of a gel of several soluble proteins in which are embedded filaments of collagen, the protein found in skin and connective tissue throughout nature (p. 43). These filaments can slide around in the gel and thus offer little resistance to mechanical stresses. But in response to a physical impulse, such as a nudge or a tap, two (or perhaps more) small cross-linking proteins are released from the innermost layer of the skin and bond the collagen filaments into thick rigid bundles to effect the transformation. Other protein factors are responsible for reversing the change, but the details are still unclear.

How then to simulate such versatility without the advantages of all those years of evolutionary exploration? This is a matter of some interest, for the attributes could be useful in fabrics responsive to a change in the weather, or in artificial muscles that would contract when a gel becomes an elastic solid. There are copolymers that will change their form but retain a memory of their designated shape. A favourite macromolecule of the futurologists is N-isopropylpolyacrylamide (NIPAAM), which has the formula:

$$
\left(\!\!\begin{array}{c} CH_2-CH \\ | \\ C=O \\ | \\ N \quad CH_3 \\ H \quad C \\ H \quad CH_3 \end{array}\!\!\right)_n
$$

It can be seen that this polymer is amphipathic (p. 167), being made up of both hydrophobic and hydrophilic elements—the branched hydrocarbon chains and the relatively polar amide groups. It sits, in fact, on a knife edge between the forms of behaviour that characterize the two extremes, and depending on conditions, one or the other will prevail. At room temperature or below it is soluble in water, but when the temperature rises its hydrophobicity comes to the fore, the bound water is released, and the polymer falls out of solution as a compact white solid. In copolymers with other monomer units this process can be finely tuned by selecting the transition temperature; or the polymer can be made pH-sensitive by incorporating some acrylic acid units into the chain:

$$
-CH_2-CH- \\ | \\ COO^-
$$

183

Under mildly acid conditions the charged carboxylate groups, $-COO^-$, will then be reversibly converted to the uncharged carboxyl $-COOH$.

If such polymers are chemically cross-linked, the soluble form is converted into a hydrogel, an insoluble swollen, watery gelatinous mass. Change the conditions to favour the hydrophobic state, and the polymer will lose its water and shrink by perhaps a hundredfold in turning itself into a dense solid. Such a huge volume change could perhaps be harnessed to power a robotic muscle. Or might such a polymer perhaps be fashioned into a weather-responsive fabric? Or if enzymes were to be coupled to the hydrated polymer, they would become largely shielded from the watery medium and from anything dissolved in it, when the gel is converted into a dehydrated solid; and whatever reaction the enzyme catalyses would then be terminated.

Polymers with a shape memory have been made by combining the building blocks of thermoplastics with those of a polar material, such as acrylic acid, as before. A copolymer of this kind will undergo a transition from a rigid to a soft and elastic state at some critical temperature. An article produced by moulding at the higher temperature, at which the material is soft, can be deformed into some other shape, into which it will set when cooled and rigidified. When warmed it will return to its original form. The imagination can be relied on to devise applications. There are even now polymers that change colour when stressed.

Supramolecular polymers

Here we encounter an entirely different kind of beast, with its own special attributes. Supramolecular polymers are by any useful definition polymers, but their monomer units are not covalently linked. Instead they associate with each other (in all cases so far devised) through hydrogen bonds (p. 24). Hydrogen bonds are much weaker than a covalent bond, but under the right conditions, and collectively, they can form tight attachments. The creation of such polymers began a little over ten years ago with the synthesis of a monomer with four hydrogen-bonding sites to generate the potential for sufficiently strong interactions. An additional design feature for supramolecular polymers that can survive in water is to shield the hydrogen-bonding sites with hydrophobic groups to protect them against the competing hydrogen-bonding action of the surrounding water (p. 25). The monomer units, then, consist of a short chain of some kind, with the quadruple hydrogen-bonding group at both ends. It looks something like this, where the shaded portion represents the short-chain.

These units simply click together in a polymer chain, as shown.

So why bother? The answer is that hydrogen bonds are highly sensitive to temperature and to the nature of the surrounding solvent. A solution of this polymer at normal temperature, at which the hydrogen bonds are strong enough to maintain the

185

structure indefinitely, will have all the usual properties, such as high viscosity. But when warmed the hydrogen bonds weaken and the polymer progressively falls apart, first into shorter chains, with a large decrease in viscosity, and then at high enough temperature into the monomer units. This is equally true of the solid polymer, which may be hard when cold but will soften in an adjustable manner when warmed—a great advantage for moulding and other processes. The wiles of the polymer chemist also allow the creation of more complex monomers, which will assemble into copolymers, block polymers, and other forms, each with its own special properties.

Supramolecular polymers are a growth area. Self-healing materials, which have only to be warmed to seal a crack, or even just squeezed, have already been made. Adhesives with special properties are another application, and also printer inks, which flow in the hot pen and set solid on the paper. There is plenty of further scope for the imagination.

The fight against emissions

Buses now plying in some of the world's more enlightened cities emit no carbon dioxide or other environmentally unfriendly effluent. They are powered by fuel cells, devices which convert chemical fuel into electricity without burning it. Without giant molecules this would scarcely be possible. In the standard fuel cell hydrogen and oxygen molecules undergo conversion at two opposite electrodes, separated from one another in the reaction chamber by a polymer membrane. At one electrode, the anode (+), each of the two atoms of the hydrogen molecules, H_2, loses its solitary electron, e, and is converted to a proton, H^+. The protons, seeking the cathode (−), migrate through the membrane. Oxygen, O_2, is supplied in the form of air, and a reaction takes place that leads to the formation of water (with no smoke): $4H^+ + e + O_2 \rightarrow 2H_2O$. The electron flow does the work of energy accumulation. The membrane is the key to the technology, for it must conduct the protons selectively without dissolving or deteriorating (for protons are, after all (p. 30), nothing more than acidity). The material in question is generally Nafion (developed by DuPont), a fluorinated (and therefore chemically stable) polymer like Teflon, but carrying acidic (negatively charged) groups. More recently a polymer of still greater durability has been synthesized, also based on a fluorocarbon. It is a copolymer of a fully fluorinated polyether (p. 160) and the acidic polystyrenesulphonic acid, with pendant groups

$$-\overset{\overset{\textstyle O}{\|}}{\underset{\underset{\textstyle O}{\|}}{S}}-O^-$$

It has the advantage that it can be moulded into any desired shape, size, and thickness.

Electrics: the advent of plastic metals

The chemistry of polymers gives, as must by now be apparent, wide scope for developing materials with new and useful properties. In electrics they now do far more than just insulate cables. Polymeric conductors, which have already been in use for some years, are among the most remarkable. Chains with conjugated backbones—those with alternating single and double bonds, in which the electron 'cloud' extends over the whole assembly of atoms (p. 157)—will conduct electricity (though by no means as effectively as, say, copper). The breakthrough came with the discovery in Japan of a method of polymerizing acetylenic monomers with ($-C\equiv C-$) bonds to generate such conjugated chains. One use, quickly found, was in the construction of batteries, but the great advantage of course is that conducting materials can now be cast into any required shape. Because of the exceptional capacity of carbon nanotubes and graphene sheets to conduct electricity, they can give rise, when incorporated into a polymer in only small proportions, to a highly conducting composite, which may in time supplant most others.

Conducting polymers have found applications in chemical sensors, in which a surface is coated with a molecule or group to which the substance to be analysed binds with high affinity. The receptor must undergo a change that induces, for instance, a rise in electrical conductivity of the matrix to which it is attached. This then results in transmission of a signal. Such sensors can have prodigious sensitivity: they are in effect a kind of synthetic nose—like a dog's nose even, which can sense a few molecules of another dog's bouquet on a lamp post. How to detect tiny concentrations of metal ions in solutions is a commercially important problem, for which metal-binding polymers have been developed. They are chemically complex, for the metal atoms are sequestered in rigid 'cages' of precisely orientated electron donor groups. One such polymer looks like this:

When the polymer with this formidable formula sees the metal ion, M, it snaps it up. The interest in this particular substance is that its electrical conductivity increases when it binds the metal, which could be highly advantageous in the design of a sensor.

Biosensors are equally important in clinical medicine, and in general depend on an enzyme that acts on the molecule to be detected, converting it into a product capable again of engendering an electrical or optical signal. Such systems play an important part in the so-called 'laboratory-on-a-chip' (of which more later), a development that has been transforming clinical and industrial chemistry. A range of other uses for conducting polymers, for instance in the construction of a synthetic nerve to conduct an electrical impulse and activate a muscle in a paralysed limb, have been proposed.

A *semiconductor* is a material that conducts electricity only under certain conditions, when it is warmed for example, or when it is irradiated with light of a specified wavelength. The properties of semiconductive materials can be manipulated by 'doping', that is, by adding small quantities of impurities. There are two kinds of doping agents: one is a source of electrons, the other a trap for electrons, or 'hole'. The first produces a p-type, the second an n-type semiconductor. In contact, the two create a p–n junction. Current (electrons) can then flow in only one direction across the junction. Diodes and transistors are constructed from assemblies of p- and n-semiconducting layers. Polymers with semiconducting properties, constructed from a conjugated backbone (most often a molecule called poly(N-vinylcarbazole), or PVK), with doping groups at desired intervals, are now commonplace.* An innovation is the flexible transistor, created from conventional

* Graphene sheets (p. 96) are also semiconductors, but of a unique kind, for they are uniform structures in which the electrons do not hop between traps, but

189

inorganic semiconductors, or from carbon nanotubes, deposited in two separate microscopically thin layers on a flexible polymer sheet. The alluring vision is of a computer or television display screen that can be rolled up and stuffed in a coat pocket.

Another recent development is the polymer liquid crystal display. Liquid crystals are fluids with some properties of a solid. The molecules of the liquid crystal may exist in the random state or can line up in an orderly array (the geometrical arrangements of the molecules depending on the nature of the substance). Such alignments generate special optical properties, which may reveal themselves in the appearance of rainbow colours. Liquid crystals occur in nature and account, for instance, for the shimmering colours in the shards of beetles. In liquid crystals formed by molecules with the tendency to cluster together in orderly fashion (of which molten cholesterol is one), the switch from the disordered to the ordered (generally helical) state can be induced by a change in temperature or, most usefully, by application of an electric field. A characteristic of an intrinsically asymmetric medium—a quartz crystal, say—is that it is optically active: it will change the plane of polarized light passing through it (p. 20). This is no less true of a liquid crystal. The problem is only that in a liquid there is motion, so ordered arrays are free to diffuse around, and are in general randomly disposed in the body of the liquid. They can all be brought into the same orientation in a thin layer if the electrodes supplying the voltage to induce order are themselves properly orientated. The liquid crystal display unit (LCD) consists of a sheet of polarizing plastic (as in the lenses of sunglasses), the cell containing the liquid crystals, another polarizing layer with

flow at constant speed, not far below the speed of light. This makes them attractive targets for fast electronic devices of the future.

its axis at right angles to, or 'crossed' with, the first, and a sheet of reflecting plastic (or a light) at the back for viewing. The windows are covered with conducting polymer layers, serving as electrodes. The polymer chains in the electrodes are all aligned, and induce uniform orientation of the liquid crystals; they are also patterned in the form of a numerical display. When there are no liquid crystals in the cell, the crossed polarizers will allow no light to pass, but with liquid crystals in the cell and the voltage switched on the plane of polarized light where it passes through the liquid crystals is rotated, so that it is no longer blocked by the 'read' polarizer; then patterns (digits) determined by the voltage appear in the display.

Polymer-based liquid crystals have been prepared by attaching to an elastomeric polymer (one with elastic properties, like rubber) a series of small light-sensitive rod-like appendages of a kind that will form conventional liquid crystals. They project like fishbones from the spine, and can cling to one another, thereby inducing alignment of the elastomers, with a resulting change in refractive index or colour. Such a system, which adjusts its optical properties or its shape very quickly (for the unconstrained elastomer chains are free to slide about, almost as in a liquid), and with an infinitesimal input of energy, affords many options beyond the construction of efficient optical displays.

The parallel between the shape change of an assemblage of polymer molecules and the contraction of a muscle has not been lost on bioengineers (a growing population in engineering departments of universities). The spectre of a surrogate synthetic muscle has appeared before them, and one path towards this is a light-activated liquid crystal elastomer. Chemical groups that change shape on absorption of light of a fixed wavelength, by virtue of

191

a switch from *trans* to *cis* about a double bond (p. 109), can be attached to the polymer chain. The light absorption of the *cis* form is different from that of the *trans*, so light of the one wavelength will cause contraction, while light of the other will reverse the process. Will this be the route to artificial muscles, whether for robots or for amputees? There are agents (actuators, as they are termed) other than light which can be made to effect changes of shape; these include addition or elimination of ions, such as calcium, and the application of a voltage. There are elastomers with charged groups that change shape in such circumstances. A sheet of such a substance is coated, top and bottom, with soft electrodes, perhaps a carbon-containing paste. On application of the voltage across the sheet, the charged groups are driven closer together as they strive towards the electrodes, and the thickness diminishes. But the volume must remain constant, and the structure therefore stretches to compensate, spreading the charges in the plane of the sheet. Very substantial (and entirely reversible) increases in area have been achieved in this way in response to a modest applied voltage. Such mechanisms are of interest in relation to robotic and other devices. A further interesting development is the invention of self-sealing rubbers. These differ from the conventional kind in that the polymers are cross-linked by numerous hydrogen bonds (p. 24), instead of covalent linkages. Their combined strength is sufficient to keep the mass intact, and they will reform spontaneously between severed surfaces. Such 'self-healing' materials could find many applications if their stability and elasticity is up to the mark.

Polymers chemically related to conducting or semiconducting types, but with magnetic properties, have also been created. One such embodies chemical groups that bind iron atoms, and thereby render the material strongly magnetic, like iron itself. It has, in fact, the properties (though not quite the strength) of an iron magnet. Applications in data storage are envisaged.

Carbon nanotubes, with semiconductive properties (p. 189), may find similar uses. Indeed, in the opinion of some electronic engineers, a revolution in the design of circuitry on a submicroscopic scale for computers (logic gates) and control units of all descriptions, through the introduction of conducting and semiconducting carbon nanotubes (p. 102), is imminent. A memory storage device has been designed based on pairs of nanotubes, one of which is given an electric charge (the 'on' state), while the other discharges it ('off' state) when it makes contact by bending.

Photonics

Photonics is a hybrid science, conceived only a few years ago. It concerns itself with such matters as chemical sources of light, new lasers, transmission of light by fibre optics and other means, electro-optical instruments, and the like. Its practical applications range from energy generation to communications and information processing. The liquid crystal displays (above) are one outcome, but synthetic and even natural polymers (proteins especially) now permeate developments in the field.

There are now many polymers that change their properties when light falls on them. The most useful to date have been photorefractive materials—those that change their refractive index* when illuminated. So a plastic sheet with this property,

---

* The refractive index of a medium (say water or glass) can be viewed as the degree that a light ray entering the medium is bent (like a pencil in a glass of water appearing kinkea).

193

irradiated through a mask in parallel stripes, will be imprinted with alternating bands of higher and lower refractive index. Light passing through such a grid will generate an interference pattern. This happens when light rays are combined, for light travels as a wave, and where the peaks of such two waves coincide the intensity will be reinforced, while where peak encounters trough the intensity vanishes:

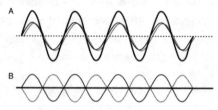

Figure 34 Interaction of light waves. A. Two rays travelling in phase reinforce one another; the result of adding the two intensities is the wave of twice the amplitude. B. Two waves out of phase interfere, and the resultant intensity is zero (bold line).

Light rays passing through a regular pattern of any kind (arrays of holes or stripes, say) will create a pattern of equal regularity on a screen. The refractive index imprinting process is reversible, for the alternation is caused only by the migration of electric charge from the shielded to the illuminated regions. This, then, amounts to an erasable memory.

An interference image in space is a *hologram*. The concept (and the word for that matter) sprang from the brain of a Hungarian-born physicist, Dennis Gabor, in 1947. He had at the time a quite different objective in mind than generating three-dimensional images in space, and he retrospectively called his insight 'an experiment in serendipity' which was 'begun too soon'. The reason was that its useful exploitation had to await the invention of the laser, which came only in 1960. (Gabor nevertheless, and very properly, eventually received the Nobel Prize for his brainwave.) The laser gave birth to an immense range of new possibilities because of the property called coherence: a coherent light beam is one in which all the rays oscillate in phase—peaks and troughs in perfect coincidence. An equally important virtue of laser light is that its wavelength distribution is exceedingly narrow: all other light sources emit a far broader wavelength spread. A hologram can be captured and displayed in a holographic medium, such as a photographic film: when viewed through the film, a three-dimensional image suspended in space comes into view. Holography has many applications, in even such everyday technologies as the reading of bar codes and the embossed three-dimensional security image on a bank card.

In the photorefractive device, stacks of polymer sheets with alternating higher and lower refractive index stripes can produce, by virtue of constructive interference (intensity reinforcement wherever two light rays oscillate in phase—the peaks of the waves exactly coinciding), a very high reflectivity for light of a chosen wavelength, determined only by the thickness of the layers. The refractive index is controlled by 'doping'—in short, contaminating—the polymer sheets with a suitable material. This is commonly a collection of tiny beads, a few ten-thousandth of a centimetre across; they are generally made of silica (quartz), but block polymers have also been synthesized, in which the blocks spontaneously fold on themselves to form submicroscopic crystal-like cubes, to produce a similar effect. With such devices holographic images can be repeatedly created, expunged, and replaced. Other geometrical forms should inspire new developments.

Then again, polymers in electro-optical devices may put out an electrical signal when illuminated by light of a particular wavelength, or they may emit light when a voltage is applied to them. They all have a conjugated (p. 187), semiconducting backbone, sometimes with special light-absorbing groups attached. A wide range of such polymers has been created, because chemical manipulation of the constituent monomers can alter the wavelength of the emitted light as required. Such polymers form the basis of organic light-emitting diodes (OLEDs). One of their many advantages over the conventional inorganic type is that that they can be deposited by spraying or printing on a hard or a soft substrate. A normal OLED is constructed in the form of a sandwich, with the photoemissive polymer as the filling between a layer of a conductive polymer and an inorganic cathode sheet. The conducting polymer layer is in contact with the anode, and the whole is encased in a box with a transparent window. When a small voltage is applied to the cell, electrons and holes (p. 189) enter the semiconductive polymer and electric charge flows through it, causing it to emit light. This type of device may serve as the basis of the next generation of flat-panel display screens. Many light-emitting polymers have been made, each radiating its characteristic colour. This means that deposition of a pattern of such polymers on a sheet of a tough, perhaps flexible, polymer by inkjet printing will be able to generate a multicolour display.

There are many variations on this design. One makes use of a semiconductive dendrimer (p. 161) with a light-emitting group shielded at its core; this offers the advantage of more efficient charge transfer. If polymers that emit light of different wavelengths are applied by inkjet printing to a polymer substrate as an array of colour-pixels, coloured images can be displayed. But for this to happen transistors must activate the individual pixels, a development yet to be achieved. 'Smart' pixels can take the form of minute fluid-filled bubbles, embedded in the plastic matrix. All contain solid electrically charged particles of dye of one or another colour which, under the influence of the external voltage generated by the transistor, migrate to the bubble's surface and thereby reveal themselves. An erasable plastic newspaper operating along such lines is an aspiration.

Printing in miniature

Lithography (from the Greek *lithos*, stone, and *graphein*, write) is a venerable engraving process, familiar enough to elicit a dramatic poem from the Victorian Scottish bard, William Topaz McGonegall, still celebrated today for his uniquely execrable verses. This opus contains the memorable lines:

> And when life's prospects may at times appear drear to ye,
> Remember Alois Senefelder, the discoverer of Lithography.

Today the technique of photolithography is a central element in microfabrication, most notably of microchips for the integrated circuits that control computers, mobile phones, and so many other accoutrements of modern life. The principle is simple enough. A mask with a pattern of holes is the template. It is placed over a silicon wafer coated with a 'photoresist', or 'resist' for short, in the form of a reactive polymer. The resist is rendered insoluble in a chemical solution only after a reaction provoked by ultraviolet light. Irradiation through the mask then prints a latent image, as in photography, on the wafer surface. The wafer is treated with the solvent (the 'developer'), which dissolves away the parts of the film that have seen no light, and an embossed pattern emerges. Then a conducting, semiconducting, or insulating layer can be applied to the available surface, and the residual resist is removed. In reality, the process is a great deal more complex and exacting than this outline suggests. In between the basic steps, a number of treatments, especially cleaning and baking at controlled temperatures, are needed to generate a flawless product. There are, besides, many variants of the technology. For instance, the above scheme is based on a 'positive' resist, but a 'negative' resist is also sometimes (though now rarely) used. In that system it is the unexposed area which is insoluble, and a 'negative' pattern is created on the wafer. In either case, the demands on reproducibility and precision are fearsome. The photolithographic technique is also now being developed for producing components based on conducting or semiconducting polymers, and for creating patterned surfaces for nanofabrication, made from light-reactive polymers. These are only a few of the polymer photonic contrivances already in use or in prospect.

Biomimetics ascendant

Biomimetics is a trend word—a neologism, which stands for the design of devices and materials inspired by the study of what evolution has contrived. The computing and printing systems based on the photosynthetic machinery of purple bacteria are one (as yet unrealized) product. 'Smart' microlenses afford a very different example of polymer-based biomimetics. Miniaturized optical systems have many applications and as many advantages. When, in the eighteenth century, anatomical dissections first revealed the structure of the eye, and Thomas Young and others began to contemplate our remarkable ability to vary our focus at will, a debate ensued about how this was managed. Young suggested that muscles in the eye could exert pressure on the lens, so as to change its curvature and thus its focal length, but the opposing theory was preferred, namely that the lens could move back and forth, like that of a camera. (Young was derided by his rivals, and it was many years before his explanation prevailed.) A 'smart' or 'tunable' lens, which can change its focus, is clearly a simpler and more robust device than anything with moving parts—with knobs and gearwheels, as in a microscope or camera. Minimizing the number of rigid moving parts will in general lead to greater economy and reliability.

This is why microlens systems have come into use in photonics, and especially in biomedical equipment. The obvious way of controlling the curvature of a lens is by pressure. The lens may consist of nothing more than a curved surface, created at the interface between water and a layer of oil floating above it. A small plate, pierced by a circular hole about 2 mm in diameter sits in the interface between the two liquids, so that the water below meets the oil above only in this aperture. The sides of the aperture are coated with a hydrophilic film, wetted by the water, whereas its upper surface is hydrophobic. The water in the aperture therefore rises to make a sort of unsupported droplet, as if on a greasy surface or on Teflon. The plate is supported on a ring of a hydrogel (a stiff watery jelly) of a temperature-sensitive polymer, resting on a solid base to make a closed chamber. When the temperature is raised, the polymer gel loses water and contracts. A pressure drop results, the plate descends a little and the curvature of the water bubble in the aperture diminishes. Other ways can of course be found to change the structure of the polymer gel,

199

which could be made sensitive to acidity or light or concentration of ions in the water. Many variants on this type of system can be envisaged.

A different route is to design a compound eye, like that of a fly. The insect eye is made up of several thousand units, called *ommatidia*, packed together on a dome-shaped surface to afford an enormous field of vision. Each ommatidium consists of a lens, attached to a cone, which collects the light and leads it along a light guide to a detector cell, called a photoreceptor. A structure of this kind—an artificial insect eye—has been made from several different types of polymer by microfabrication techniques—in minute moulds, with microfluidics and light- or heat-induced polymerization, and where necessary cross-linking of the polymer constituents. Each artificial ommatidium is no more than one-fifth of a millimetre in length, and a fifth of that in diameter at the business end. This tiny polymer lens collects the light, which passes down a polymer cone into a light guide. The light guide is a plastic rod, about $1 \mu m$ (one-hundredth of a millimetre) across, with a jacket, consisting of a different polymer of lower refractive index to prevent the light from escaping. At the end of the rod is the photodetector. There is of course far more to the design and fabrication of such a device, which is still in the development stage, but many applications, inside the body for clinical investigations in particular, are proposed.

The ingenious nocturnal moth evades its predators with the aid of an anti-reflection surface in the back of its eyes, which appear black, rather than glowing like the eyes of other insects, when a light shines on them. The moth achieves this by dint of a regular array of small conical protuberances, about 200 nm (two ten-thousandths of a millimetre) in size, with a refractive index that changes from tip to base. The reflection of light from glass, say, is a consequence of the abrupt refractive index change which the light rays encounter when they strike the surface. The discovery of the moth's secret instantly inspired attempts to create artificial anti-reflective surfaces. Success was achieved by using transparent polymers, such as polystyrene, to fabricate minuscule bubbles with a polymer concentration, and therefore refractive index, which varies from top to bottom. The proportion of sunlight reflected from a surface studded with these bodies is less than 2%. This principle has also been adopted for the construction of solar cells, the conventional surfaces of which reflect—thus waste— 30–40% of the incident radiation. Related methods are used to reduce the reflection of microwave (radar) radiation in stealth aircraft, and elsewhere.

Microarrays and microlaboratories

Microarrays of DNA, are large ordered patterns of micro-scopic spots of DNA pieces—often tens of thousands of them— deposited on a solid surface, a centimetre or two across, commonly referred to as a genome chip. The grid is formed by deposition with a needle attached to a robot arm, or through holes in a photolithographic mask. For most applications, the spots are either derived from genes in the actual genome (the totality of the DNA sequence in our cells), or are synthetic DNA pieces (oligonucleotides) with sequences corresponding to a chunk of the gene sequence (in practice most often twenty to thirty nucleotides (see p. 88) long). To make the DNA corresponding to genes, the messenger RNA from cells has to be purified. This RNA corresponds to the genes actually expressed, in other words those giving rise to the protein which they

201

encode (p. 93); the bulk of the DNA, which contains no genes, is never translated into messenger RNA. The messenger RNA can then be copied back, with the aid of an enzyme, into a corresponding stretch of DNA (called complementary DNA, or cDNA).

These microarrays have become highly important tools in research and in the clinic. Suppose, for example, you want to find out whether a particular gene is expressed in a type of cell that you are studying. You extract the RNA from cells of that kind, recovered from tissue or cultured in the laboratory. The RNA will include the messenger for your gene if that gene is being expressed. You then attach a fluorescent marker to the collection of RNA molecules by a simple one-step chemical reaction. You irrigate the chip with this solution, and allow each messenger RNA to attach by base-pairing (making a double helix) to its complementary DNA. You wash away unattached RNA, and examine the chip in a robotic reader, which produces an image of the grid with the fluorescent spots where the messenger has bound. Then where your gene fragment sits in the array there will be a fluorescent spot—or not. With oligonucleotide chips a hospital laboratory can determine whether a patient or a newborn infant has a functional or defective gene. The pharmaceutical industry can identify subjects with a particular genetic make-up (genotype), to see whether the action of a drug is related to that genotype. And so on. There are now a number of variants of microarray technology, among them arrays of proteins instead of DNA.

Economical and accurate microarray analyses—assays of enzymes for instance—can be incorporated into mini-laboratories—the 'lab-on-a-chip'. Nanotechnology has given birth to many versions of this device. All the ones now in currency have dimensions of a few square millimetres, and are designed to perform a variety of chemical and physical operations. In such a laboratory, analytical procedures are carried out on solution volumes in the picolitre range, that is, a few thousandths of a billionth of a litre, or some billionths of a cubic centimetre. Manipulations on this scale are made possible once again by microfluidics, circuits of pipes, pumps, and valves, produced by microfabrication of polymers (most commonly the silicon-based poly(dimethoxysilane)). The virtues of such systems are numerous, for once the means of manufacture have been developed, they can be made in limitless quantities for very little cost. They are fast in response and economical of materials, which may be precious or sparse—often the case with clinical or forensic samples—or dangerous in bulk, if toxic or highly radioactive. Several chips of this kind can be connected in parallel to perform linked operations, such as the large range of tests that go into a comprehensive blood analysis for haemoglobin, serum proteins, fats, salts, hormones, and other constituents. The lab-on-a-chip is also leading to a cheaper way of sequencing whole genomes, for instance of pathogenic bacteria or viruses, so that different strains can be identified. Plant and animal genome sequences are of enormous importance for selective cultivation, and of course genetic engineering.

A protein computer

A scheme for faster computing takes its inspiration from something close to the working of the retina. The purple bacterium, *Halobacterium halobium*, thrives in hostile environments, such as

203

the Dead Sea and salt marshes, with their enormous salt concentrations, often at a very high temperatures—conditions lethal to other living organisms. Oxygen supply is uncertain and so the bacterium has evolved a unique mechanism for obtaining its energy: it is photosynthetic, like a plant, but its energy receptor is not chlorophyll. It possesses instead a protein closely resembling the rhodopsin, on which mammalian vision depends (p. 59). It is called, in fact, bacteriorhodopsin and contains essentially the same conjugated vitamin A-related molecule as our own 'visual purple'. And in just the same way it absorbs light, undergoes a structural change (*cis–trans* switch about a carbon–carbon double bond), which bleaches its colour and sets off the sequence of metablic reactions involved in energy storage. In actuality, the configurational switch that follows absorption of a photon by both bacterial and mammalian rhodopsins proceeds through several steps, all completed within a few milliseconds (thousandths of a second). But when the protein is cooled down, the process slows and the normally short-lived states can be captured. The idea then is to confine the bacteriorhodopsin in a cooled transparent matrix. The intermediates all have somewhat different colours, and can be interconverted by irradiation with laser light of the appropriate wavelength. This, in a pixel array, could serve as a memory chip with a switching speed perhaps a million times faster than a transistor can manage. One laser would effect the transition between two of the states, while a second, emitting light of the colour that the protein in its altered state now absorbs, would deliver the readout. More elaborate devices based on the options afforded by the series of intermediates in the bacteriorhodopsin photocycle (as it is termed) have been conceived. It is not clear whether this line of inquiry really has a future in computing. DNA, which we will come to, is perceived by most computer engineers as a better bet.

Microengines

Photolithography is applied, in conjunction with cutting by an atomic force microscope, an instrument that (in one of its modes) stabs a sharp tip into a substrate, to fabricate minuscule mechanical components, such as gear wheels and levers. These are used for constructing nanomachines (as we must now call them). The idea is to use these for further microfabrication, and also in surgery and for other clinical purposes. But the smallest motor yet has been constructed on a truly microscopic scale by physicists in California, and makes use of a multiwalled carbon nanotube, no more than 10 nm in diameter, as a drive shaft to propel a gold rotor. The whole contrivance is 500 nm in size—some three-hundred times smaller than a human hair—and the rotor spins at 2000 or more rpm. An attribute of the carbon nanotube that invites exploitation is the slippery nature of the graphene surface (p. 96). This means that in, for instance, a double-walled nanotube, the inner cylinder can be made to slide out of its housing at minimal energy cost. The scope of devices that will take advantage of this type of telescopic movement is being explored.

The field of technology known as nanorobotics, which concerns the creation of machines operating on the microscopic or submicroscopic scale, is beyond the scope of this discourse; in fact it would require a book to itself. *Nanobots*, as they are generally called, will, if you believe the zealots, eventually control our everyday lives. The American engineer and visionary Eric Drexler asserted some twenty years ago that the time would soon come when nanobots, based perhaps on motor proteins, working in shoals, would be programmed to assemble anything from small to huge domestic and industrial machines. Others foresee nanobots performing almost any service we might desire, from cleaning our environment to curing our diseases and healing our wounds. A less extravagant ambition is to develop nanobot sensors for monitoring metabolites in the body. So when, for example, the glucose level in the blood of a diabetic exceeds the desired limits, a nanobot will activate an implanted radiofrequency transmitter, and an alarm will sound, a telephone ring, or an insistent voice call, for an insulin injection. Or of course another nanobot will turn the tap and release the precisely metered insulin dose.

Drexler, it should be said, warned that the day might come when nanobots would acquire the capacity to reproduce (presumably by parthenogenesis), and escape from the control of their creators; there would then be no telling what mayhem might ensue. (He did not, though, advert to the more mundane threat that clouds of nanoparticles might present to health—the 'grey goo' spectre that occasionally grips the imagination of journalists.) Another American engineer, Hans Moravec, presents a more benign scenario: thousands of nanobots will be able to work through the neurones in the human brain, reading off all the information stored in them (no matter that the nature of memory storage is still far from clear). The information will be downloaded onto disc, or whatever ultrahigh-density storage device is by then in current use; this could then be installed in a mechanical device of your choice, which will stimulate on demand whatever pleasure centres in its/your silicon brain it/you might wish, or access all the world's intellectual or cultural resources. Backup copies will be stored underground, in satellites, or on heavenly bodies other than ours, and then your immortality will be assured.

But enough! To return to the present: an already active line of endeavour is to harness the known protein motors—still a lot smaller than the smallest man-made contrivance—to do useful work on a molecular scale. Myosin, the motile protein of muscles, is one that we have already met (p. 63), but myosins of many specialized kinds engage in a variety of other physiological activities, from growth and replication of cells to working the sound detection mechanism in the ear. Another motor protein, kinesin (p. 65), contains, like muscle myosin, two large identical subunits. The central parts of their sequences are α-helices, twisted together to form a coiled-coil. Two globular domains project at either end; those at one end (the 'feet') are the motor units. They travel, one step of 8 nm (billionths of a metre) at a time, along a protein track, called a microtubule. (The myosins, on the other hand, travel along a track of the different protein, actin.) Both kinesin and myosin use ATP (p. 62 f.n.) as their fuel. The kinesin works in the cell as a cargo transporter. The load is attached to the globular 'heads' at the site of production and delivered to the place where it is needed.

These properties have exercised the imaginations of nanotechnologists, for if the microtubule tracks can be laid on a surface, the kinesin molecules can be despatched with their cargoes to nano-scale factories—assembly plants perhaps for nanomachines. Rudimentary transport systems based on kinesin have been constructed, but work on a different principle, for the mode of kinesin action in nature can be usefully stood on its head: rows of kinesin molecules can be stuck to a prepared surface, ruled for instance with an atomic force microscope, their feet uppermost. Microtubules can then be made to glide along the kinesin tracks. (Recall Einstein's purported inquiry of a guard at Paddington station: 'Does Oxford stop at this train?') Chemistry would be required to attach the cargoes to the microtubules, and allow for their release at the terminus. The train could also be switched on or off if a photochemically activated derivative of ATP is substituted for simple ATP: an ultraviolet light flash then releases ATP from its 'caged' state and sets the machinery in motion. The type of application envisaged for machines driven by kinesin or other motor proteins is in the regeneration of damaged tissue in the heart or nervous system, or for building the next higher size of nanomachines.

Rotary motions also occur in nature, and nanotechnologists have been eying the proteins that drive them. There are the motors that turn the flagella of bacteria (p. 65), and there is a remarkable protein in the membranes of mitochondria (p. 91) called the F_0F_1 ATPase. This protein has the form of a rotating drum, made up of eight segments on a stalk, which projects through the membrane and acts as a channel for ions. It is a fully reversible motor, using the energy inherent in a gradient of ions (in this case hydrogen ions, or protons, H^+ (p. 30)) to make ATP from ADP (p. 63). It takes energy to keep the protons in a region in which their concentration is high and prevents them from dispersing uniformly in the water (just as it takes energy to keep a heavy load suspended above the ground). The drum rotates in the anticlockwise direction, moving in increments of one segment, spewing out an ATP molecule at each step, while protons flow through the stalk and annul the gradient. When there is no proton gradient left, the drum starts to rotate clockwise, powering the motion by converting the ATP to ADP, while pumping unwanted protons out through the stalk. A larger rotor could also be constructed from the kinesin-microtubule system with a circular track.

10

THE MANY FACES OF DNA

DNA, known to most of us only as the genetic material, the storehouse of heredity, has re-emerged in the past two decades in all manner of new guises. For it embodies in its chemical nature a remarkable range of technological options, not all related to the storage of information.

DNA machines

Nanotechnology has been concerned, as we have seen, with the construction of miniature motors. Some laboratories are now striving to develop gadgets on biomimetic principles to drive a rotor with the aid of a ratchet-and-pawl device on an atomic scale—and in fact chemists have synthesized machines, based only on small organic molecules, to rotate or perform other motions when an energy source, sometimes in the form of light, is supplied. Such simple creations may find applications as switches, but can generate no useful force. A simple torsion-generating device can be made from circular double-helical DNA, which has the capacity to form a supercoil. Tightening the twist of the primary double helix of a circular DNA induces supercoiling, as described on p. 92. The result is a positive supercoil, in which the primary helix is said to be overwound. Conversely, slightly untwisting the primary helix of the circle causes supercoiling in the opposite sense (negative supercoiling when the primary helix is underwound). Supercoiling can be induced in a circular DNA by special enzymes with the ability to nick, then twist, and finally reseal the primary helix. The amount of superhelical twist in the DNA loop can then be manipulated by any external agency that changes, ever so slightly, the natural double helix pitch. Changing the concentration of salts in the solution is one way; another is by intercalation. Many flat molecules containing fused rings (p. 17) have an affinity for the DNA bases, and if added to the DNA solution will slide (or intercalate) between adjacent bases in the double helix. When this happens the helix must untwist a little to accommodate the intruders. The screw increases its pitch length, lowering the tension induced by the intercalation, and the supercoils relax. If the circular DNA is not initially supercoiled intercalation induces negative supercoiling, because the loop now must accommodate a greater length of primary helix.

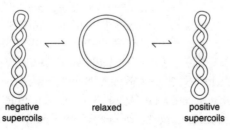

negative supercoils   relaxed   positive supercoils

Figure 35 Supercoiling of a closed circular DNA double helix.

So by altering the composition of the medium a torque can be induced in the DNA, in either direction and reversibly, and the system can be made to do work.

Many other ideas about how the remarkable versatility of the DNA structure could be exploited have—in principle, at least—been realized. So, for instance, a closed circular DNA with two contiguous stretches of sequence complementary to each other can form a hairpin, projecting from the straight double helix. With two such identical regions in the sequence a cruciform structure results:

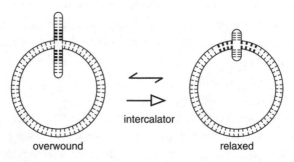

overwound   relaxed

Figure 36 A simple engine made from a closed circular DNA, in which adjacent parts of the sequence are complementary to each other.

If now an intercalating agent relaxes the primary supercoiled circle, some of the base pairs near the cruciform junction will move into the circle, and the projections will contract (and vice versa when the intercalator is removed). So here again is a rudimentary engine.

Yet another has been devised, based on a switch from a right- to a left-handed helix (B- to Z-form of DNA), which can be brought about merely by the addition of different ions (magnesium, for example). Considering the variety of proteins that interact with DNA—which break the chain or seal the breaks, replicate the sequence, twist or untwist the helix, or even bend it at a sharp angle when they bind—there seems to be limitless scope for imagination in devising DNA-based contrivances on the Meccano principle. DNA machines that stretch, twist, rotate, and even walk with two single-stranded legs have been described, though none have yet been put to practical use. Whether Drexler is right when he insists that this kind of device foreshadows mighty technologies of the future is probably for the next generation to discover.

DNA as enzyme and sensor

In 1994 an unexpected discovery was made in an American laboratory: DNA is an enzyme. This ought perhaps not to have come as such a surprise, since a form of its close relative, RNA, had been found a decade earlier to have an enzymic activity (p. 93): it could break molecules resembling itself into smaller pieces. This observation, truly unexpected at the time, gave rise to the notion of the 'RNA world', a new slant on the ever-vexed question of how life originated on Earth. According to this theory, RNA with its, perhaps multiple, enzymic properties was the primal, self-replicating molecule. It could transform itself into messenger and transfer RNAs, which would then assemble the amino acids swimming in the primeval soup into proteins, and then it would be only a few billion years before man strode the Earth. The discovery of enzymic DNA changed that picture, but, whereas self-digestion of RNA plays an essential role in the cell, no function has yet been found for the same activity in DNA.

The fact, though, is that it is being put to good use in the laboratory, for it has been found that the reaction requires the presence of a metal ion, and the identity of that metal ion depends on the structure of the DNA. Neither double- nor single-stranded

DNA will do the trick, which needs instead a more complex form, not seen in nature, containing some double helix with a loop at one end, and a bulged-out loop elsewhere in the sequence. Its preferred substrate is actually RNA, which the DNA can break much more rapidly than another DNA chain. It also turns out that that fracture of the chain always occurs in a double helix formed between the substrate (the RNA) and a complementary sequence in the DNA. The picture below shows the structure of such a DNA enzyme (also called a DNAzyme), and the way it acts:

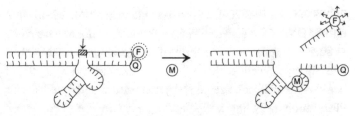

Figure 37 Structure and action of a DNA enzyme. M is the metal ion needed for enzymic activity, F is the fluorescent group and Q the quencher, which extinguishes the fluorescence when it is close to F. The shaded bar is the bond which is broken by the enzyme function.

The rules relating the structure of the DNA to the particular metal ion it requires for its catalytic action are still obscure, and so to find a match a 'combinatorial' method of synthesis must be used. Such methods play a major part in modern chemistry. They amount to constructing mixtures of trillions of sequences by letting the starting materials combine repeatedly at random. When applied to the synthesis of DNA from its four nucleotides, A, C, G, and T, a mixture of chains, containing an astronomical number of different sequences, all around, say, fifty nucleotides in length, is generated. Any one of these with a binding specificity for a particular metal is picked out by exposing the mixture to that metal, attached to a solid support. All that then remains is to amplify the piece of DNA by polymerase chain reaction (PCR) (see p. 224). Such enzymic DNAs, specific for lead, uranium, and other metal ions, have been made, and incorporated in a sensor for detecting the metal and measuring its concentration in drinking water. The sensor works like this: a dormant fluorescent group is attached to the part of the chain that breaks off when the metal ion activates the enzymic activity (as in the picture above). It can easily then be arranged for the fluorescence to reveal itself only when the fragment breaks free; this is done by attaching a group called a quencher close to the source of the fluorescence. High sensitivities of detection have been achieved with devices built on this principle.

The DNAzyme has also been fashioned into a nanomachine, opening and shutting like tweezers. Two double helices, acting as jaws, are joined by a short single-strand segment, which is flexible and so allows the legs to open and shut freely. In addition, sequences at the open ends are made complementary to the ends of a DNAzyme of appropriate length, which accordingly binds between the tweezers' jaws. This particular DNAzyme is a simple single-stranded piece of DNA which allows the tweezers to remain in the relaxed, closed position. The substrate, with a sequence complementary to the main part of the DNAzyme, surrounds the machine in the solution. A substrate molecule binds to make a stiff double helix between the jaws and drives them apart. The enzyme breaks the substrate into two short pieces, which float away into the solution, and the jaws again close. In this way the device opens and shuts, until the substrate—its fuel—is exhausted. The inventors have even added a brake, in the form of a complementary DNA that displaces, or at any rate competes with the substrate, and slows down the action.

DNA tiles and Lego

A new focus of nanotechnology is the production of patterned surfaces and three-dimensional structures on a molecular scale. But to what end such investment of effort and ingenuity? One answer is simply that, as an article of faith, all new technologies will eventually bring unforeseen returns. Another, less evasive, is that a patterned surface will serve as a base for the construction of three-dimensional structures on a molecular scale for performing mechanical operations (machines), or perhaps for high-density information storage. The great physicist Richard Feynman offered a prize for the first nano-device that could inscribe the contents of the *Encyclopaedia Britannica* on the head of a pin. The challenge was not taken up in his lifetime, but something close to what Feynman envisaged has now come to pass. We have seen that appropriately designed block polymers can generate regular patterns, but it is manipulation of DNA that has thrown up the most spectacular results.

The point about DNA is first that defined sequences of a hundred or more nucleotides can be chemically synthesized, replicated by PCR (p. 224), and elongated by joining several of these identical pieces end-to-end with an enzyme. Secondly, double-stranded DNA can be produced in this way, with a projecting single-stranded segment at one end, to which another DNA with a single-stranded end complementary to the first can be attached by hybridization (formation of a length of double helix). Such

segments are termed 'sticky ends'. One can also arrange for
strands from adjacent identical double helices to crossover and
thereby link the two molecules in a rather rigid fashion:

Figure 38 Making a link between
two DNA double helices by
uniting (hybridizing)
complementary stretches of
sequence.

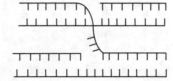

Such, effectively cross-linked, DNA molecules will form stiffish
parallel arrays, and if properly designed sticky ends can be put in
place, these units ('tiles') will self-assemble on a surface to which
DNA can adhere to form a carpet.

Another way is to start with single-stranded DNA and use
short pieces (oligonucleotides) with sequences in tandem, complementary to those of opposite chain ends as cross over linkers to
assemble the rig. By using the crossover technique, together with
sticky ends located at selected positions, it has proved possible
to construct elaborate shapes—stars, triangles, and even 'smiley'
faces, all on the scale of a few hundred nanometres (some ten-
thousandths of a millimetre, and far too small to be seen under a
light microscope). A field of these, revealed by the atomic force
microscope, is a disconcerting apparition: *

---

* The top row of Figure 39 shows how the pieces of DNA, designed to generate
the shapes through the association of sticky ends, assemble to create the designated
shapes. The second row shows the arrangement of DNA helices crossing. The third
row shows images of single complexes, seen under the atomic force microscope (an
instrument with resolving power similar to that of the electron microscope). The
bottom row shows fields of such particles. The objects on the far left are larger
than the others. The white bar on that picture is 1 μm in length (1 micron, or one
ten-thousandth of a centimetre); the bars on the other pictures are 100 nm (one
hundred-thousandth of a centimetre).

217

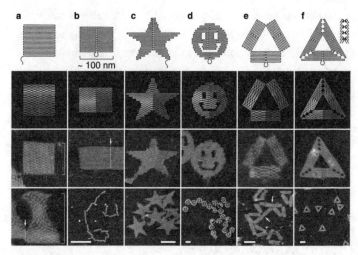

Figure 39 Self-assembled DNA patterns.

Such simultaneous assembly of all components is something that happens in nature but not on the factory floor.

The same method has been used to create three-dimensional structures, such as tetrahedrons (p. 19), dodecahedrons (symmetrical hollow bodies with twelve pentagonal faces), and footballs, like buckminsterfullerene (p. 97). The way these are built up from small DNA components is shown below. Flexibility, needed to make a figure in three dimensions, is generated by introducing short single-stranded loops of unpaired nucleotides.

The remarkable feature of these forms is again that they assemble spontaneously in a mixture of the three starting materials, just like large natural complexes, such as multi-subunit proteins, or ribosomes or viruses. Which of the three-dimensional structures shown here preferentially forms depends on the concentration of

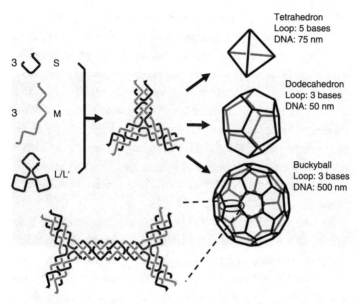

Figure 40 Self-assembly of symmetrical structures from pieces of DNA. On the left are the synthesized pieces of DNA, which spontaneously assemble into the forms on the right, through the intermediate forms in the centre. The final shape is determined by the concentration of the components in the solution and the lengths of the small single-stranded bits needed to provide flexibility. These are the triangular element in the middle of the piece marked 'L/L', which appear in the middle of the upper figure in the centre and where the forks diverge in the lower figure.

the components: if it is low the tetrahedrons are favoured, if high the larger bodies. (Keep in mind that in the closed shell all binding sites (sticky ends) are satisfied, so one form will not convert itself into another.) The DNA dodecahedrons are not very rigid, and can be squeezed without breaking.

There are many variations on this theme. The laboratories that operate in this area now have a sort of toolbox of girders, struts, rivets, staples, and four-way junctions, all made of DNA, from which all manner of structures can be assembled. The design of components becomes rapidly harder the more complex the desired architecture, and computer programmes have been developed to assist. A further level of design involves the incorporation of active groups, such as those that bind a metal atom by chemical coupling through the nucleotide sugar (deoxyribose) on a chosen component of the rig. DNA-binding proteins may also be allowed to attach themselves at nucleotide sequences specific for such interactions. DNA boxes with lids that open have also been made.

And so we come again to the question, 'But what use is it?' When, about 150 years ago, Michael Faraday demonstrated his discovery of electromagnetism to William Gladstone, the Chancellor of the Exchequer, he is said to have answered the very same question thus: 'I don't know, but one day, sir, you may be able to tax it.' At this stage one can only conjecture, but considering again the many proteins that recognize and attach to a unique sequence of nucleotides, one can envisage building up complex structures to do chemistry or mechanical work. A more immediate application is submicroscopic electrical circuitry. By choosing the DNA sequences sites can be created for the attachment of metal atoms, such as gold. Because the geometry of the DNA double helix is so precisely defined, the metal atoms can be deposited in a highly regular pattern, a huge advantage for the construction of nanoelectronic circuitry. Such metal assemblies on DNA matrices may come into their own as information storage devices, and also as sensors and highly efficient catalysts for chemical and biological reactions. The development of such a remarkable variety of DNA-based structures is, at all events, already a remarkable and spectacular achievement.

The end of the silicon age? Computing with DNA

To date the development of information storage systems has adhered to Moore's Law of computer technology, the principle enunciated by the computer guru Gordon Moore that storage capacity doubles every eighteen months. But the storage capacity of the silicon computer chip will ultimately reach a limit when miniaturization can go no further. There are, it seems, computer visionaries who believe that the answer lies in DNA, for its storage capacity is truly immense (p. 94), as the study of genetics has taught us. And DNA is cheap. A kilogram of the stuff contains more information in its four-letter code than all the computers on the planet.

Storage, though, is all very well, but how can the information be written and retrieved? An exploratory experiment, based, to be sure, on laborious biochemistry, was performed by L. M. Adleman more than a decade ago. As a first demonstration of the watery DNA computer in action he chose to solve a simple form of what mathematicians know as the Hamiltonian path, or 'travelling salesman' problem—how, starting from city A, to travel to each of a succession of other cities, terminating at city Z, without ever retracing your path. A formal solution to this ancient mathematical teaser was worked out in the early nineteenth century by two mathematicians, Sir William Rowan Hamilton in Ireland and Thomas Kirkman in England. With their aid an answer can be easily found if the number of cities is small, but if there are many it requires an enormous amount of computer time.

Adleman solved the problem for a small (seven-city) set using only the standard techniques of DNA manipulation. He assigned to each of the seven cities an oligonucleotide sequence (the order of bases, A, C, T, and G, in a short strand of DNA), and to the flight-path connecting any pair of cities another sequence. The flight-path sequences were made up of the last half of the sequence signature of the city of departure and the first half of the city of destination signature. So suppose city **A** is designated by the sequence CCGAT<u>ATCGG</u>, and city **B** by <u>ATAGC</u>GGAAC, then the flight path **A–B** becomes ATCGGATAGC. Adleman synthesized the fourteen oligonucleotides defining all possible connecting flights (or to be precise, he ordered them from a commercial company, as most molecular biologists do these days), and in addition his system called for seven oligonucleotides *complementary* to those designating each of the seven cities. In other words, for city **A** he procured the complementary sequence GGC-TATAGCC, for city **B** the complementary sequence TATCGC-CTTG, and so on. The complementary sequence would form a double helix by base-pairing A to T and G to C with the signature sequence of the city, but the signature oligonucleotides are not themselves needed for the present.

So now the test tube computation can begin. Something like 1 μg (millionth of a gram) of each of the collection of twenty-two oligonucleotides (containing some 10^{14} molecules) are mixed together in solution, and an enzyme that joins breaks in a DNA double helix, called a ligase, is added. What happens now? Complementary sequences find each other in the solution and click together in a double helix. These products will include, for instance, the **A–B** flight-path sequence and the **B** city complementary sequence, which will bind to one another to make a partial double helix:

$$\begin{array}{c} \text{ATCGGATAGC} \\ \vdots\ \vdots\ \vdots\ \vdots\ \vdots \\ \text{TATCGCCTTG} \end{array}$$

Projecting from this five-base-pair double helix is a 'sticky end', ready to take up the flight-path sequence **B–C**, since the run of five bases on the right will be complementary to its first five nucleotides, and we have:

```
ATCGGATAGC • GGAACNNNNN
: : : : :   : : : : :
     TATCG  CCTTG
```

where NNNNN are the last five nucleotides of the **B–C** flight-path, and at the same time the first five of the city **C** signature. The dot in the upper strand shows that there is a break in the chain, which the ligase immediately seals.

In this way representations of all possible journeys are created in the test tube, and the problem now is to extract the correct answer from the innumerable incorrect ones that start and finish in the wrong cities. First of all, the newly synthesized DNAs are analysed by gel electrophoresis, the procedure originally used in sequence analysis: the DNA mixture is applied to a transparent polymer gel, through the pores of which the DNA can travel under the influence of an electric field. Since it is negatively charged it will move towards the positive electrode, the anode, and the tangle of polymer chains making up the gel acts as a filter, slowing down the longer DNAs more than the shorter. The migration distances can be calibrated with DNA samples of known lengths, and so the DNA of the right length (in this instance, corresponding to six flights) can be identified and recovered. Then the method of hybridization—of creating hybrid double helices by association of complementary DNA strands—is used to discard those components of the mixture that do not contain all seven cities. An oligonucleotide of a given sequence will attach to a DNA containing the complementary sequence, and if tethered to a solid support, will pull it out of the solution. All seven city oligonucleotides are applied in turn, so one finishes up with the DNA mixture containing all seven cities (and the size tells one that each city is present only once). And now one uses the oligonucleotides with the short sequences that will only bind to the two terminal sequences of the end-product if these are city **A** and city **G**—the start and the end of the journey. The PCR is then used to copy thousands of times over (amplify) the desired sequence and nothing else. The technique is inseparable from the practice of molecular biology and genetics. It has revolutionized forensic science, for one can now discover whether the DNA of any individual has a counterpart in the minutest trace of semen or blood, whether two people are related or what the protein make-up is of a mummified Egyptian. The point is that PCR will deliver you quantities of DNA sufficient for any kind of analyses from just a few molecules. But now a digression.

The details of PCR need not concern us, but here, for the inquisitive, is a brief description, illustrated below. (1) We start with a double-stranded DNA, with a sequence we wish to repli-cate in bulk (amplify). (2) The two chains run (p. 89) in opposite directions, as indicated. We prepare (or order, more likely) two short oligonucleotides, with sequences complementary to the ends of the segment of DNA we wish to amplify, one in the posi-tive strand (the gene or other sequence of interest), the other in the negative (complementary) strand. We add a generous quantity of these *primers*, as they are called, to the DNA, heat the solution to separate the DNA strands, and allow the single strands to anneal to their primers on cooling. (3) Then we introduce the enzyme, DNA polymerase, which copies the DNA strands, starting only from the primer (because it operates only from double-helical

DNA), to make complementary chains of variable length. The new chain contains the sequence complementary to the opposite primer. (**4**) We heat to release the new products, cool, again allow the primers to bind, and the polymerase to extend the chains, starting again from the attached primer. (**5**) We repeat the process, but now the chain stops at the integrated primer, and since the polymerase can copy no more than what it sees, we get only the sequence that we originally defined. (**6**, **7**) After that, successive cycles of heating and attachment of more primer molecules will result in only more of the same products, and the amount of the amplified DNA produced will rapidly increase with each round. The desired product (bottom left panel) accumulates in a manner analogous to that perceived by the wily sage, invited by the Chinese emperor to name his price for services rendered: the emperor thinks he has a bargain when the sage asks for two grains of rice to be placed on the first square of a chess board, $2 \times 2 = 4$ grains on the next, $4 \times 4 = 16$ grains on the one following, and so on. The 64th and last square would then have had to accommodate the entire rice crop of the empire. The PCR process is automated, and wholly routine. An important feature is the use of a polymerase enzyme which will not itself be destroyed (denatured) by the heating step, and this derives from one of the thermophilic bacteria, which thrive at temperatures close to boiling water, and have heat-resistant proteins (p. 67).

To return now to the travelling salesman—Adleman got the right answer, though of course, this is a preposterously labour-intensive form of computation. But it does illustrate the unique power of DNA as a device for retention and manipulation of information. Even in Adleman's rudimentary procedure it shows its capacity for parallel processing, far beyond anything that the most advanced supercomputers can encompass, since many

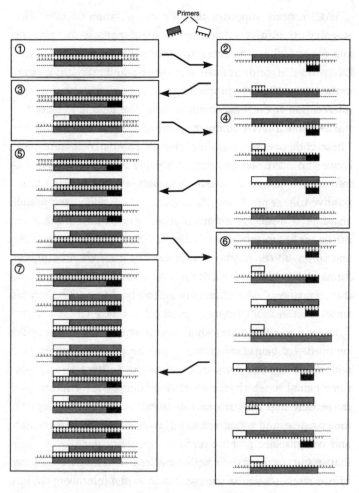

Figure 41 The polymerase chain reaction (PCR).

billions of binding and ligation operations are simultaneously performed. The economy, in terms of energy requirement, is also incomparably superior to that of silicon computers.

Well, nobody supposes that the computation of the future is going to involve laboratory manipulations with test-tubes and pipettes. Since Adleman's pioneering demonstration, many DNA-based algorithms have been devised and followed through in the laboratory, even proceeding to completion without human intervention at every biochemical step. The essential element in the design of a DNA computer is the construction of logic gates. These components, consisting in your computer of transistors, convert an electrical input signal—voltage on or voltage off, say—into binary, the fundamental language of the computer. So a positive voltage produces unit output 1, or *yes*, and a zero voltage an output 0, or *no*. The output, or those of several such logic gates, can be applied to another logic gate in the next level of complexity, and so on, all the way to a network. Logic gates come in three forms, designated AND, OR, and AND-NOT, which constitute the components of a circuit. In the DNA computer the input signal is a short DNA strand, instead of a voltage.

Keep in mind that very short double helices are unstable: the centipede hanging on with one or two legs has a weak grip, but, with all its legs deployed, its hold is tight. In the case of DNA, three paired bases are the absolute minimum for a meaningful interaction, and the strength rises rapidly as more are added. A long single strand will attach tightly to its complementary strand, and will displace a short piece of complementary DNA from that strand. In the picture below we start with a DNA made up of two strands with partial complementarity, joined together by their complementary segments. DNA 1 is a single strand with two sequence segments, which we will call A^- and A_t^-, where t stands for toehold, for reasons which will become obvious. This is the input to the logic gate, which contains (1) a strand composed of sequence, A^+, complementary to A^-, and A_t^+ complementary to the toehold, A_t^-, and (2) a second strand comprising A^-, B^-, and B_t^-. Strand (1) is tethered to a solid support, while strand (2) becomes the output, when displaced by the input strand, which can form a much longer, therefore stronger duplex with the tethered logic gate strand. The output strand, with the composition $A^- B^- B_t^-$, serves as a *yes* signal. In the scheme depicted it becomes the input to a second logic gate with a tethered strand and a weakly bound output strand, as before. Displacement leads to a second and final *yes* signal, $B^- C^- C_t^-$.

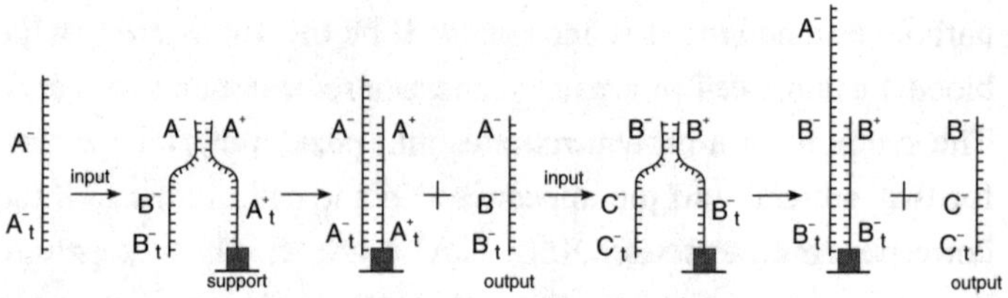

Figure 42 A DNA logic-gate arrangement.

In much the same way more complex networks of logic gates can be constructed—requiring, for instance, the simultaneous input of two or more different DNA strands—and we have the makings of a computer. The molecules do not swim around in the solution, since the permanent component of each unit logic gate is chemically tethered to a solid support. This is only one of many elaborate schemes for creating DNA-based systems of logic gates. Others involve the joining of separate DNA strands by an enzyme (ligase), and their scission by a different enzyme, a so-called restriction nuclease, which cuts the chain at the same point. Many laboratories are engaged in the pursuit of a DNA computer, each of them developing its own system. All aim at creating a DNA chip, assisted probably by microfluidic circuitry (p. 203). But

228

the problem remains that both DNA hybridization and enzymic reactions are very much slower than electronic switching. So far as can be foreseen therefore, DNA-based computers, assuming they come to fruition, will be confined to a restricted range of applications. But as the Danish physicist Niels Bohr was reputed to have remarked, 'Prediction is very difficult, especially of the future.'

The type of reasoning underlying the design of logic gates is also being applied to the formulation of circuits for DNA-dependent sensors or automata. Suppose, for example, that a pathological condition is accompanied by the appearance in the bloodstream or cell of a particular protein (as is often the case). The synthesis of a protein requires the transcription of the gene for that protein, and the appearance in the cell therefore of the corresponding messenger RNA. The sequence of the gene and thus of the messenger (which is single-stranded) being known, a complementary DNA strand can be made. A shorter complementary strand can then be displaced from the DNA by the messenger, and becomes the input to a logic gate. This would amount to a simple two-state circuit: is the messenger present, yes or no? An output nucleotide could perhaps activate an enzyme to destroy the messenger and shut off synthesis of the protein, or release a drug. But this still seems far removed from us in time. Yet, several automata based on the lab-on-a-chip principle are no doubt already on the drawing-board.

The versatility and power implicit in the structure of DNA shows yet again what the blind process of long evolutionary tinkering has achieved. Science will find no end of uses for biological molecules—the proteins and nucleic acids and their complexes, tested by nature, unrelenting and merciless, over the aeons. As one of the leading thinkers in the field has remarked, 'Engineers would be foolish to ignore the lessons of a billion years of evolution.' At the same time, few experts now hold that DNA really will supplant silicon in the foreseeable future: there is, on the one hand, the limitations on speed, and on the other, the vast quantities of DNAs of different sequences that (it seems at this stage) would be needed to make a working computer with power higher than that of the current generation of supercomputers. Yet the intellectual allure of the field is undeniable. There have been many theoretical advances and promise enough to keep computer engineers and nanotechnological visionaries engaged in the contemplation of DNA far into the future.

Further Reading

Polymers

R. J. Young and P. A. Lovell, *Introduction to Polymers*, 2nd edn (Chapman and Hall, 1991).

M. P. Stevens, *Polymer Chemistry* (Oxford University Press, 1990).

C. Tanford, *Physical Chemistry of Macromolecules* (Wiley, 1961).
Old as it is, this book remains the best of its kind.

Biological molecules and their functions

B. Alberts, A. Johnson, P. Walter, J. Lewis, M. Raff, and K. Roberts, *Molecular Biology of the Cell*, 5th edn (Taylor and Francis, 2008).

H. Lodish, A. Berk, C. A. Kaiser, and M. Krieger, *Molecular Cell Biology*, 7th edn (Scientific American Books, 2007).

J. M. Berg, J. L. Tynocko, and L. Stryer, *Biochemistry*, 5th edn (Freeman, 2006).

Proteins

M. F. Perutz, *Protein Structure* (Freeman, 1992).
A lucid survey by one of the great figures in the field of the way in which protein structures are determined, and their bearing on medical research.

C. Branden and J. Tooze, *Introduction to Protein Structure*, 2nd edn (Garland, 1998).
A beautifully illustrated compendium of the structures and structural motifs of the various families of proteins.

Nucleic acids

V. A. Bloomfield, D. M. Crothers, and I. Tinoco, *Nucleic Acids: Structure, Properties and Function* (University Science Books, 2000).

Materials

R. W. Cahn, *The Coming of Materials Science* (Pergamon, 2001).

J. E. Gordon, *The New Science of Strong Materials—Why You Don't Fall through the Floor*, with a new introduction by P. Ball (Princeton University Press, 2006).

An updated classic for the lay reader.

P. Ball, *Made to Measure—New Materials for the 21st Century* (Princeton University Press, 1997).

An excellent general account of developments in materials and their practical applications.

General and Historical

Y. Furukawa, *Inventing Polymer Science—Staudinger, Carothers, and the Emergence of Macromolecular Chemistry* (University of Pennsylvania Press, 1998).

C. Tanford and J. A. Reynolds, *Nature's Robots—A History of Proteins* (Oxford University Press, 2001).

An absorbing history of the subject from its dawn to its high noon.

U. Lagerkvist, *DNA Pioneers and their Legacy* (Yale University Press, 1998).

An amiable account for the lay reader.

H. F. Judson, *The Eighth Day of Creation: The Revolution in Biology*, 2nd edn (Cold Spring Harbor Laboratory Press, 1996).

This is a highly recommended and compelling history of the emergence of molecular biology.

Bibliography

Chapter 1

J. T. Edsall, Proteins as macromolecules: an essay on the development of the macromolecule concept and some of its vicissitudes. *Arch. Biochem. Biophys.* **Suppl. 1** (1962), 12–20.

H. Morawetz, *Polymers: The Origins and Growth of a Concept* (Wiley, 1985).

R. Olby, The macromolecule concept. *J. Chem. Ed.* **47** (1970), 168–74.

C. Tanford and J. Reynolds, How protein chemists bypassed the colloid/macromolecule debate. *Ambix* **46** (1999), 33–51.

And see also C. Tanford and J. A. Reynolds, *Nature's Robots—A History of Proteins* (Oxford University Press, 2001).

Chapter 2

P. W. Atkins, *The Elements of Physical Chemistry*, 4th edn (Oxford University Press, 2005).

W. H. Brown, *Introduction to Organic Chemistry*, 3rd edn (Wiley, 2003).

R. J. Dubos, R. Dubos, and T. D. Brock, *Pasteur and Modern Science* (ASM Press, 1998).

K. Mislow, *Introduction to Stereochemistry* (Dover, 2003).

Chapter 3

Protein structure

B. Alberts, A. Johnson, P. Walter, J. Lewis, M. Raff, and K. Roberts, *Molecular Biology of the Cell*, 5th edn (Taylor and Francis, 2008).

C. Branden and J. Tooze, *Introduction to Protein Structure*, 2nd edn (Garland, 1998).

J. M. Berg, J. L. Tynocko, and L. Stryer, *Biochemistry*, 5th edn (Freeman, 2006).

M. F. Perutz, *Protein Structure* (Freeman, 1992).

M. Vendruscolo and C. M. Dobson, A glimpse at the organization of the protein universe. *Proc Natl Acad Sci USA* **102** (2005), 5641–2.

Mutations and disease

K. E. Davies and A. P. Rees, *Molecular Basis of Inherited Disease* (IRL Press, 1992).

D. M. Swallow and Y. H. Edwards (eds), *Protein Dysfunction in Human Genetic Disease* (Bios, 1997).

Folding and misfolding

C. M. Dobson, Protein folding and misfolding. *Nature* **426** (2003), 884–90.

C. M. Dobson, Prying into prions. *Nature* **435** (2005), 747–9.

Y. Ivarsson *et al.*, Mechanisms of protein folding. *Eur. Biophys. J.* **37** (2008), 721–8.

Fibrous proteins

C. Cohen and D. A. D. Parry, α-Helical coiled coils and bundles: how to design an α-helical protein. *Proteins* **7** (1990), 1–15.

A. W. Kajava, J. M. Squire, and D. A. D. Parry, β-Structure in fibrous proteins. *Adv. Protein Chem* **73** (2006), 1–15.

M. Van der Rest and R. Garrone, Collagen family of proteins. *FASEB J* **5** (1991), 2814–23.

Elastic proteins

J. Gosline *et al.*, Elastic proteins: biological roles and elastic properties. *Phil Trans Roy Soc B* **357** (2002), 121–32.

M. Rothschild and J. Schlein, The jumping mechanism of *Xenopsylla cheopis*. 1. Exoskeletal structure and musculature. *Phil Trans Roy Soc B* **271** (1975), 457–90.

This is one of the many publications by the incomparable Miriam Rothschiled, who made a life's study of the habits of fleas and lice. (See the popular book by Miriam Rothschild and Theresa Clay, *Fleas, Flukes and Cuckoos: A Study of Bird Parasites* (Collins, 1952, reprinted by Arrow (1961).)

Spider and other silks

C. Dicko, J. M. Kenney, and F. Vollrath, β-Silks: enhancing and controlling aggregation. *Adv Protein Chem* **73** (2006), 17–53.

O. Emile, A. Le Floch, and F. Vollrath, Shape memory in spider draglines. *Nature* **440** (2006), 421.

F. Vollrath, Spider silk: thousands of nano-filaments and dollops of sticky glue. *Curr Biol* **16** (2006), R925–7.

Composites

P. B. Messersmith, Multitasking in tissues and materials. *Science* **319** (2008), 1767–8.

C. Ortiz and M. C. Boyce, Bioinspired structural materials. *Science* **319** (2008), 1053–4.

S. A. Wainwright *et al.*, *Mechanical Design in Organisms* (Princeton University Press, 1976).

Enzymes

N. C. Pace, *Fundamentals of Enzymology: The Cell and Molecular Biology of Catalytic Proteins* (Oxford University Press, 2000).

Chapter 4

Sugars and polysaccharides

D. A. Rees, *Polysaccharide Shapes*, Outline Studies in Biology (Chapman and Hall, 1977).

Glycoproteins and proteoglycans

L. Kjellén and U. Lindahl, Proteoglycans: structures and interactions. *Ann Rev Biochem* **60** (1991), 443–75.

W. J. Lennarz (ed.), *The Biochemistry of Glycoproteins and Proteoglycans* (Plenum, 1980).

R. J. Woods, Three-dimensional structures of oligosaccharides. *Curr Opin Struct Biol* **5** (1995), 591–8.

Nucleic acids

R. L. P. Adams, J. P. Knowles, and D. P. Leader, *The Biochemistry of the Nucleic Acids* (Chapman and Hall, 1992).

And see especially the books by Alberts *et al.*, Lodish *et al.*, and Stryer *et al.* (above).

W. Gilbert, The RNA world. *Nature* **319** (1986), 618.

J. D. Watson and F. H. C. Crick, A structure for deoxyribose nucleic acid. *Nature* **171** (1953), 737–8.

This is the original paper in which the double-helical structure of DNA was put forward. It is probably the most cited paper in the history of science, and a historical landmark, well worth a visit.

Chapter 5

Buckminsterfullerene (the buckyball) and graphene

H. Aldersey-Williams, *The Most Beautiful Molecule: The Discovery of the Buckyball* (Wiley, 1997).

A. K. Geim and K. S. Novoselov, The rise of graphene. *Nature Mater* **6** (2007), 183–91.

P. J. F. Harris, *Carbon Nanotubes and Related Structures—New Materials for the 21st Century* (Cambridge University Press, 1999).

D. Li and R. B. Kaner, Graphene-based materials. *Science* **320** (2008), 1170–1.

M. Reibold *et al.*, Carbon nanotubes in an ancient Damascus sabre. *Nature* **444** (2006), 286.

Chapter 6

Plastics

H.-G. Elias, *An Introduction to Plastics* (Wiley, 2003).

R. Friedel, *Pioneer Plastic—The Making and Selling of Celluloid* (University of Wisconsin Press, 1983).

I. Fuyuno, Plastic promises. *Nature* **446** (2007), 715.

See also P. Ball, *Made to Measure—New Materials for the 21st Century* (Princeton University Press, 1997).

J. I. Meikle, *American Plastic—A Cultural History* (Rutgers University Press, 1995).

A learned and entertaining history of plastics in modern life.

Chapter 7

Polymer science

P. J. Flory, *Principles of Polymer Chemistry* (Cornell University Press, 1953). This remains the classic at the graduate level—not for the lay reader.

See also R. J. Young and P. A. Lovell, *Introduction to Polymers*, 2nd edn (Chapman and Hall, 1991); and Y. Furukawa, *Inventing Polymer Science—Staudinger, Carothers, and the Emergence of Macromolecular Chemistry* (University of Pennsylvania Press, 1998)—for a historical treatment.

Rubber

J. Loadman, *Tears of the Tree: The Story of Rubber—A Modern Marvel* (Oxford University Press, 2005).

But for the hard science of rubber elasticity, see P. J. Flory, *Principles of Polymer Chemistry* (Cornell University Press, 1953).

Properties of materials

R. E. Hummel, *Understanding Materials Science: History, Properties, Applications*, 2nd edn (Springer, 2004).

J. F. V. Vincent, *Structural Biomaterials* (Princeton University Press, 1990).

See also J. E. Gordon, *The New Science of Strong Materials—Why You Don't Fall through the Floor*, with a new introduction by P. Ball (Princeton University Press, 2006).

Composites

P. J. Hogg, Composites in armor. *Science* **314** (2006), 1100–10.

A. Miserez *et al.*, The transition from stiff to compliant materials in squid beak. *Science* **319** (2008), 1816–19.

P. Ungar, Strong teeth, strong seeds. *Nature* **452** (2008), 703–5.

J. H. Waite, Adhesion a la moule. *Integr Comp Biol* **42** (2002), 1172–80.

And see again J. E. Gordon, *The New Science of Strong Materials—Why You Don't Fall through the Floor*, with a new introduction by P. Ball (Princeton University Press, 2006).

Chapter 8
Stereoregular polymers—the history

F. M. McMillan, *The Chain Straighteners* (Macmillan, 1979).

Dendrimers

S. Hecht and J. M. Fréchet, Dendrimer encapsulation of function: applying nature's site isolation principle from biomimetics to materials science. *Angew Chem Internat Ed* **40** (2001), 74–91.

D. A. Tomalia, Dendrimer molecules. *Sci Am* **272** (May 1995), 62–6.

Chapter 9
Adhesives and coatings

W. J. P. Barnes, Biomimetic solutions to sticky problems. *Science* **318** (2007), 203–4.

A. K. Geim *et al.*, Microfabricated adhesive mimicking gecko foot-hair. *Nature Mater* **2** (2003), 461–3.

H. J. Hektor and K. Scholtmeijer, Hydrophobins: proteins with potential. *Curr Opin Biotechnol* **16** (2005), 434–9.

H. Lee, B. P. Lee, and P. B. Messersmith, A reversible wet/dry adhesive inspired by mussels and geckos. *Nature* **448** (2007), 338–41.

T. P. Russell and H. C. Kim, Tack—a sticky subject. *Science* **285** (1999), 1219–20.

Water purification and desalination

M. A. Shannon *et al.*, Science and technology for water purification in the coming decades. *Nature* **452** (2008), 301–10.

A. Taubert, Controlling water transport through artificial polymer/protein hybrid membranes. *Proc Natl Acad Sci USA* **104** (2007), 20, 643–4.

Polymersomes and drug delivery

D. E. Discher and F. Ahmed, Polymersomes. *Ann Rev Biomed Eng* **8** (2006), 323–41.

Tissue engineering

S. Ashley, Artificial muscles. *Sci Am* **289**:4 (2003), 53–9.

R. S. Langer and J. P. Vacanti, Tissue engineering. *Sci Am* **280**:4 (1999), 86–9.

Smart polymers

B. Jeong and A. Gutowska, Lessons from nature: stimuli-responsive polymers and their biomedical applications. *Trends Biotechnol* **20** (2002), 305–11.

J. L. Mynar and T. Aida, The gift of healing. *Nature* **451** (2008), 895–6.

M. Sarikaya, Biomimetics: Materials fabrication through biology. *Proc Natl Acad Sci USA* **96** (1999), 14, 183–5.

Supramolecular polymers

P. Cordier *et al.*, Self-healing and thermoreversible rubber from supramolecular assembly. *Nature* **451** (2008), 977–80.

T. F. A. de Greef and E. W. Meijer, Supramolecular polymers. *Nature* **453** (2008), 171–3.

Fuel cells

R. F. Service, New polymer may rev up the output of fuel cells used to power cars *Science* **312** (2006), 35.

and see also P. Ball, *Made to Measure—New Materials for the 21st Century* (Princeton University Press, 1997).

Spontaneous assembly

J. V. Barth, G. Constantini, and K. Kern, Engineering atomic and molecular nanostructures at surfaces. *Nature* **437** (2005), 671–9.

D. G. Bucknall and H. L. Anderson, Polymers get organized. *Science* **302** (2003), 1904–5.

T. P. Lodge, A unique platform for materials design. *Science* **321** (2008), 50–1.

R. A. Segalman, Directing self-assembly toward perfection. *Science* **321** (2008), 919–20.

Conducting polymers

J. C. Scott, Conducting polymers: from novel science to new technology. *Science* **278** (1997), 2071–2.

P. Yam, Plastics get wired. *Sci Am* Special Fall issue (1997), 90–6.

Light-emitting and refracting polymers

R. H. Friend *et al.*, Electroluminscence in conjugated polymers. *Nature* **397** (1999), 121–7.

A. Kraft, A. C. Grimsdale, and A. B. Holmes, Electroluminescent conjugated polymers—seeing polymers in a new light. *Angew Chem Internat Ed* **37** (1998), 402–28.

O. Ostroverkhova and W. E. Moerner, Organic photorefractives: mechanisms, materials, and applications. *Chem Rev* **104** (2004), 3267–314.

Nanotechnology

L. Dong *et al.*, Adaptive liquid microlenses activated by stimuli-responsive hydrogels. *Nature* **442** (2006), 551–4.

A. M. Fennimore *et al.*, Rotational activator based on carbon nanotubes. *Nature* **424** (2003), 408–10.

H. Hess, Toward devices powered by biomolecular motors. *Science* **312** (2006), 860–1.

P. Kim and C. M. Lieber, Nanotube nanotweezers. *Science* **286** (1999), 2148–50.

S. R. Quake and A. Scherer, From micro- to nanofabrication with soft materials. *Science* **290** (2000), 1536–9.

T. Sargent, *The Dance of the Molecules: How Nanotechnology is Changing our Lives* (Thunder's Mouth Press, 2006).

M. Sarikaya *et al.*, Molecular biomimetics: nanotechnology through biology. *Nature Mater* **2** (2003), 577–85.

Chapter 10
Nanotechnology with DNA

N. C. Seeman, Nanotechnology and the double helix. *Sci Am* **290**:6 (2004), 64–75.
This is an excellent lucid survey of what has been achieved and where it is leading.

DNA machines

N. C. Seeman, From genes to machines: DNA nanomechanical devices. *Trends Biochem Sci* **30** (2005), 119–25.

DNA as enzyme

R. Bar-Ziv, DNA circuits get up to speed. *Science* **318** (2007), 1078–9.

J. Liu *et al.*, A catalytic beacon sensor for uranium with parts-per-trillion sensitivity and millionfold selectivity. *Proc Natl Acad Sci USA* **104** (2007), 2056–61.

D. Y. Zhang *et al.*, Engineering entropy-driven reactions and networks catalyzed by DNA. *Science* **318** (2007), 1121–5.

Structures from DNA

F. A. Aldaye, A. L. Palmer, and H. F. Sleiman, Assembling materials with DNA as a guide. *Science* **321** (2008), 1795–9.

Y. He *et al.*, Hierarchical self-assembly of DNA into symmetric supramolecular polyhedra. *Nature* **452** (2008), 198–201.

P. W. K. Rothenmund, Folding DNA to create nanoscale shapes and patterns. *Nature* **440** (2006), 297–302.

L. M. Smith, The manifold faces of DNA. *Nature* **440** (2006), 283–4.

P. Yin *et al.*, Programming biomolecular self-assembly pathways. *Nature* **451** (2008), 318–22.

Computing with DNA

M. Amos, *Genesis Machines: The New Science of Biocomputing* (Atlantic Books, 2006).

A popular book on the general subject of computing with biological molecules.

M. R. Adelman, Molecules as computing devices. *Scientific American* **279**:2 (1998), 34–41.

This article describes an instructive and entertaining exercise in test tube computing.

B. M. Frezza, S. L. Cockroft, and M. R. Ghadiri, Modular multi-level circuits from immobilized DNA-based logic gates. *J Amer Chem Soc* **129** (2007), 14875–9.

J. Macdonald, D. Stefanovic, and M. N. Stojanovic, DNA computers for work and play. *Scientific American* **299**:5 (2008), 84–91.

E. Shapiro and Y. Benenson, Bringing DNA computers to life. *Sci Am* **294**:5 (2006), 44–51.

Index